Crystal Radio:
History, Fundamentals, and Design

Crystal Radio:
History, Fundamentals, and Design

P. A. Kinzie

The Xtal Set Society
www.midnightscience.com
1-800-927-1771

Printed in Lawrence, Kansas
ISBN: 1-887736-05-0

CONTENTS

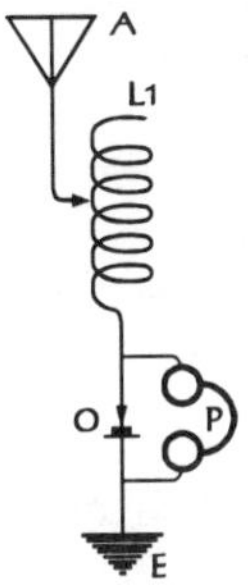

CHAPTER 1

THE HISTORY OF THE CRYSTAL SET

Introduction

The crystal set came into existence because of the invention of just one component, the detector. Everything else was easily adapted from the parts of existing circuitry. It was the development of this one key item that made the crystal set a reality. Therefore the history of the crystal detector itself is important in order to cover crystal set development. In the following sections first we cover the discovery of solid-state rectification, then the early efforts to utilize that discovery, the first practical applications, and finally, the evolution of both detector and receiver up to today's mass-produced solid-state diode.

This process took time. Some 70 years elapsed between the first successful laboratory work demonstrating the property of rectification in solids, and the first practical solid-state diodes. Surprisingly enough, relatively few people were involved to any great degree over much of this period. It was not until about the time of World War II that large scale applied research and development took place. All of these aspects of the development process are covered in the following sections.

The Discovery of Solid-State Rectification

The latter half of the 19th century was a time when scientific discoveries were made at a rapid rate, and in many cases laboratory results accumulated far faster than successful theories to explain them. This was the situation with studies of electrical conduction in solids, where theory and experiment had established only two methods of conduction, by Ohm's Law or by electrolysis.

A young German physicist named Ferdinand Braun was the first investigator to present unexpected but conclusive experimental results to the contrary. These proved that another form of conduction exists, and that the property of rectification, sometimes called unilateral conductivity, can be demonstrated in a number of natural and artificially produced solids. This experimental work was done many years before there was adequate theory to explain the results.

Braun was born in Fulda, a town in what is now Germany, on June 6, 1850. His aptitude for experimental physics and his teaching skills were already evident by the time he was in his twenties, and he later became a Professor of Physics. In the early 1870's Braun began investigating electrical conduction in certain solids that appeared to have interesting properties. He chose to test a group of natural and laboratory prepared metal sulfides, and he observed that current flow in a number of these compounds obeyed neither Ohm's Law nor the law of electrolysis.

Braun's method of mounting his specimens was unique for that time, and it turned out to be the key factor in producing his experimental results. His procedure consisted of first placing the irregularly shaped sample in a ring of silver wire. Then a second silver wire, crimped to produce spring action, was installed to press down on the top of the sample by contact at its pointed end. The result was a point contact junction, which we now know is a requirement for efficient rectification of current in the case of a large class of substances. Earlier experimenters had clamped their specimens between metal plates, or had used some similar method which exposed relatively large areas on each side of the sample to conductive surfaces. With that technique and with the particular substances tested, the results had been inconclusive at best.

Braun's results were quite unexpected; specimens connected by the new method often conducted electricity much better in one current flow direction than the other. He published his initial results in 1874, which were at first controversial. Some of Braun's colleagues disagreed with him while others reported similar results. Braun responded with three more papers on the subject, the last one being nine years later in 1883. By that time the phenomenon was fully established; there were many crystalline substances known to act as rectifiers, having the ability to conduct better in one direction than in the other.

After this period, interest in crystal rectifiers faded. A successful theoretical explanation of unilateral conductivity was years away in the far future, so there was no incentive to plan further experiments, much less a hint of what direction to take. In addition, there was not even a clue of a practical application of the effect. Finally, there were so many promising developments taking place in other fields that investigators quite naturally turned elsewhere.

Braun's final paper preceded the demonstration of electromagnetic wave propagation in space by Heinrich Hertz in 1888, and was years ahead of Guglielmo Marconi's successful wireless transmission experiments in 1895. Therefore, no thought was given by anyone about the possibility of using a crystal rectifier as a detector of electromagnetic waves. It was also true that Braun, like Hertz, had no thought of commercializing any research result that might have practical applications. The prevalent attitude of most scientists was (and in many cases remains to this day) that the results of scientific research at the universities should be available to all. In the case of Braun, however, this attitude would later change when he became involved in the development of wireless.

Solid-state Detection: the Efforts that Failed or Were Forgotten

After the publicity following Marconi's initial successes with wireless transmission, a number of companies in several countries entered the new field. Germany was no exception, where several small companies were soon competing with each other, as well as with an already powerful British corporation under Marconi's direction. Braun was swept into this rapidly developing situation in the late 1890's, when German businessmen realized the need for technical support if they were to remain competitive. He soon found himself involved in what by modern terminology is called research and development in private industry. Braun's interest and attitude changed from those of the laboratory scientist to those of the industrial inventor-developer. He experimented with wireless transmission and took active part in field tests of equipment in hopes of extending its range and improving its reliability. In fact, Braun became a major contributor to the development of wireless telegraphy, so much so that he later shared a Nobel Prize with Marconi for achievement in developing practical wireless communication. All this was in addition to work on his most well known invention, the cathode ray tube.

Braun, as well as some of the other principals in wireless development, gradually became aware of the disadvantages (and the advantage) of Marconi's original detector, an adaptation of the Branley coherer. It's big advantage depended upon the initial concept of what the ideal wireless communications system would provide. This concept envisioned the automatic recording of received signals by means of a Morse Inker or similar instrument. This eliminated the presumed disadvantage of a human operator in the receiving system.

The commitment to automatic recording resulted in a delay in accepting improved detection methods, because in general such methods were not adaptable to the automatic process. This commitment affected the Marconi organization and the German companies, including Braun's. Nevertheless, Braun came to realize the significance of his earlier work and began an effort to adapt the crystal rectifier for use as a detector in 1899. His experimental work during the 1899-1901 period was successful, and he was able to demonstrate a working detector, but incompatibility with the Morse Inker blocked its application. Unlike Marconi, Braun was a major participant but not the controlling interest in his company, and he was compelled by prevailing corporate policy to bide his time.

Meanwhile, others had become aware of the significance of solid-state rectification. The earliest of the investigators was Jagadis Chunder Bose, who is known primarily for his work in the field of plant physiology, but who also repeated the work of Heinrich Hertz, with improvements in technique and apparatus.

Bose was born in India in 1858, and received his higher education at Christ College, Cambridge England, before returning to Calcutta as a Professor of Physics at Presidency

College. He later founded the Bose Research Institute for the study of plant and insect life. He was one of the first lecturers to demonstrate Hertzian waves as a possible medium of communications, both verbally and by demonstration, using apparatus of his own design, and he suggested a number of possibilities for the use of Hertzian waves.

Bose eventually lost interest in the subject, but not before progressing far enough to recognize the usefulness of solid-state rectification as a means of detection. This is evidenced by his filing a U.S. patent application titled Detector for Electrical Disturbances, on September 30, 1901. The patent was not issued until March 29, 1904 as No. 755,840, but the early filing date indicates that he was active at about the same time that Braun was unsuccessfully trying to get approval of his own crystal detector.

The apparatus described in Bose's patent was not primarily intended as a radio frequency detector, but it was described as being readily adaptable for that purpose. The crystal detector was described in general terms as a substance possessing certain electrical characteristics. It was later determined from other information sources that Bose had actually used two dissimilar crystals of galena, in contact with each other to provide rectification. Had he fully covered this part of his work in the patent, he would have anticipated the concept of a crystal detector invention of that type which was not patented until several years later. However, his invention did meet the requirements for a practical device, although he did not produce one. Thus it was left to others to take that step.

The first solid-state detector to be useful in receiving apparatus was patented by Sidney G. Brown. This was filed June 3,

1904, just a few months before the Bose patent was issued. The detector was actually a solid-state electrolytic detector, but its macroscopic characteristics were so similar to those of a crystal detector that it was later catalogued with the latter for sales purposes. The active substance was a pellet of compressed lead peroxide powder, held between a platinum plate and a blunt lead point.

This detector was used for a time in at least one commercial wireless station in England, and an Electro Importing Wireless Catalog of 1918 was still advertising the pellets, along with crystals for detectors, at $0.30 each. It was a true invention in every sense of the word, yet it never became popular and did not change the direction of technological development. The most likely reason for the lack of success was simply that it did not have the sensitivity associated with either a liquid type of detector or the later crystal detector.

Louis Winslow Austin was another scientist who investigated rectification, and was the last to enter the scene before the burst of activity that resulted in successful detectors. He was born in Orwell, Vermont, in 1867, and attended universities both in the United States and abroad to earn his doctorate, become an Assistant Professor, and do research. He became associated with the United States National Bureau of Standards in 1904, beginning a long career that led to a number of contributions to radio science.

In the early 1900's Austin discovered that some materials which were thermoelectrically sensitive also acted as rectifiers. He incorrectly believed that rectification was caused by the thermoelectric effect, but this was understandable in view of the limited knowledge of the time. He was not alone either; a substantial number of scientists considered thermo-electric effect as a likely explanation.

Pursuing his ideas, Austin developed a detector consisting of a roller of silicon or aluminum, held under slight pressure against the surface of a block of tellurium. The tellurium available at that time was relatively impure, and its high thermoelectric power was a beneficial effect of the impurities. Austin very likely chose tellurium because of that high thermoelectric power, and he named his invention a thermo detector. He later revised his opinion when experimental results became available which showed that thermal effects were not the cause of rectification.

Austin was with the Bureau of Standards in 1906, when he applied for two detector patents, one for silicon against tellurium, the other for aluminum against the same element. Whether he was hoping for direct involvement with commercial applications, or was simply protecting his interest in the invention is not known. The fact that both British and United States patents were issued suggests that he attached considerable importance to his work. However, once again there was no further activity resulting in practical applications, at least in the United States. It should be noted however, that in Germany a similar detector called the Bronc Cell came into wide usage. The latter consisted of a finely adjusted contact between tellurium and graphite.

Solid-State Detection: the Prelude to Success

Braun, Bose, Brown, and Austin were all active at about the same time that business decisions and technology had the effect of delaying further development. Two conditions were instrumental in producing that effect and were mentioned earlier: the desire for automatic recording, and a commitment to the coherer type of detector. The bias in favor of automatic recording was particularly prevalent in Europe and England, although the United States Navy was also a proponent of that method for a time. In the latter case it has been suggested that such a system was believed to offer the advantages of less training for recruits, and fewer security problems. In any case, this favoritism naturally discouraged developments of detectors which were not compatible with automatic recording, and the crystal detector was one of these.

The coherer met the comparability requirement for application with automatic recording. It was also an attractive choice for another reason. There was an erroneous but prevalent opinion that the coherer was a device of high sensitivity, perhaps the most sensitive method of detection to be found. This opinion was even employed to excuse some of it's shortcomings, such as being difficult to adjust, unstable, and notorious for its inability to distinguish signals from static. In other words, it was considered to be temperamental because of its supposedly high sensitivity, rather than its poor characteristics.

The reason this bizarre concept was not quickly disproved, before it delayed the progress of invention and committed manufacturers to the wrong product, arose from the fact that with 1900's technology there was no way of directly measuring how much signal energy was available after the coherer rectified the incoming transmission. The fraction of signal power available at the antenna that was actually used within the coherer served to create a low resistance path for local battery power to drive a recorder pen or ring a bell. This local power was measurable, but was unrelated to the strength of the signal power. In fact, the only way that the sensitivity of the coherer could be determined was by comparison of the performance of other detectors under like conditions. This took much time and effort, mostly from the cumulative experience of a number of investigators, as well as wireless operators in the field.

A feeling that something better could be found eventually prevailed, especially in the United States, where business conditions encouraged a change of opinion. This occurred because the wireless business community in the U.S. consisted of several small companies struggling to survive in a marketplace which included German manufacturers, the American Marconi Company, and its powerful parent corporation in England. The patents held by these companies discouraged outright copying, and there were already several cases of litigation over claims of patent infringement by more subtle means. On the other hand, the bigger companies, already established, were thoroughly committed to the products based on those patents, and were understandably reluctant to change. In short, the big companies, by their own commitments, opened up some possibilities for the small ones. This was a strong incentive for the invention of new products, and for business investment to develop them.

Automatic recording was never strongly favored in the United States, but when it was realized that the presence of the human operator at an appropriate type of receiver produced a more sensitive receiving system, effectively increasing the range of transmission, the advantage was obvious. This was the way a small company could gain the edge over entrenched competition, and it created an ideal climate for American inventors. Braun and Bose could not directly benefit from this situation, being located in other nations, and Brown and Austin fared no better, even though Brown had connections in both England and the United States, and Austin was active at the Bureau of Standards. It remained for two other men to produce the detectors which came into widespread use.

The First Practical Crystal Detectors

Crystal detectors were not the first detectors to outperform the coherer, and their reign of supremacy was brief. Nevertheless, for a time they were the most used means of detecting spark transmissions, and spark was the only way there was to transmit over great distances in the early 1900's. Crystals were the best replacements for the coherer, and the person most responsible for their development was Greenleaf Whittier Pickard, who is one of the two men usually credited with the invention.

Pickard was born in 1877 and received his higher education at Harvard and the Massachusetts Institute of Technology. In 1901 he went to work for the American Wireless Telegraph and Telephone Company, which was the first wireless company formed in the United States. After some months of employment, he was working with a microphone detector during reception of signals from one of the company's transmitters. At this time the microphone detector was one of the few types not covered by a patent, and thus available to the smaller companies.

The detector was a contact-type unit consisting of carbon against steel, in series with batteries and headphones. When operated at full current, an audible signal was accompanied by a background of static, commonly described as a frying noise. While Pickard was working with this equipment, he became annoyed one day at the noise. He removed two of the three batteries in the circuit (so he thought) which eliminated the noise, although signal strength was greatly reduced. However, upon checking his circuit, he discovered that all three batteries had been removed, and that the received signal was due solely to the transmitter power received. Such reception was contrary to all of his previous experience, and it raised his curiosity.

Pickard's interest led him to experiment further with loose carbon in contact with steel, then with oxidized steel in contact with pointed steel. The oxide layer was fragile, which led him to consider using iron oxide in bulk form. In October 1902, he tried a sample of magnetite (Fe_2O_3) and received signals with better volume than any of his previous arrangements. In short, Pickard was now working with a crystal detector, discovered and tested independently of the previous investigators. During the next three years, he tested a large variety of specimens, including chalcopyrites, iron pyrites, and galena, all of which were later used in successful receivers.

By that time there was evidently a general awareness of the possibility of mineral applications as solid-state detectors. Pickard's notes show that he was acquainted with the theory that detection was a thermoelectric effect, and this influenced his choice of minerals for testing. Nevertheless, at that time and even in later years, he was still unaware of Braun's development of a practical detector.

Pickard continued his search for better materials, and in 1906 he was able to procure some fused silicon, which was at that time a new product of the electric furnace. Silicon specimens worked so well, using a point contact connection, that he was ready to patent his discovery and manufacture crystal detectors as a commercial product.

At about this time the second successful inventor entered the detector development field. Henry Harrison Chase Dunwoody, born 1842, had graduated from the West Point Military Academy in 1886. During his army career he was promoted to Brigadier General, with most of his service in the Signal Corps. His resulting experience in Army communications may have been a factor in his joining the de Forest Wireless Telegraph Company as a Vice President, after retiring from military service.

In 1906, Dunwoody was actively involved in developing a crystal detector, perhaps upon a suggestion by de Forest, and on March 23, 1906 he applied for a patent. This came at a fortunate moment for the de Forest interests, because an electrolytic detector that de Forest was using in his commercial receivers was the subject of a patent infringement lawsuit. The litigation had been in progress for some three years, and the court finally ruled against de Forest in 1906. De Forest desperately needed another type of detector when Dunwoody made his discovery.

Dunwoody took advantage of the characteristics of carborundum (silicon carbide, SiC) crystals, another product of the electric furnace. A typical batch of carborundum ranges all the way from highly colored to dull gray specimens, depending upon their location in the furnace, and he found the less colorful crystals to be the most promising. Dunwoody picked dull gray as preferable. Others have referred to greenish gray as the best choice. Dunwoody also found another property of carborundum to be helpful. Elongated crystals are not uncommon and he chose these for his specimens because electrical connections could be easily made at each end.

His method of making connections consisted of winding a few turns of wire around the ends. Current rectification occurred when there was a difference in the semiconductor properties from end to end, which occurred in a small number of the specimens. It is evident from this that Dunwoody was unaware of the advantage of using a point contact at one connection. Therefore, he either was unacquainted with or did not understand earlier results in this regard. His own results with appropriate specimens were still impressive, so he filed his patent application on March 2, 1906, preceding Pickard's application by about five months.

It is noteworthy that Austin filed for his patent on October 27, 1906, about two months after Pickard's filing date of August 30, and that is not all. Braun, returning to the scene after several years, filed for a German patent that same year. What caused this sudden burst of activity, from which Pickard and Dunwoody were to emerge with the benefits?

First, there was the growing awareness of the need for better detectors and it appears that by 1906 there was at least some general speculation that minerals could be used in this respect, as witness the interest in thermoelectric theory as well as other possibilities in explaining the phenomenon of solid-state rectification. Could this have encouraged Austin's patent application at that particular moment? The answer is not known at this time.

Pickard's situation is more evident. By all accounts he was at the stage where it made good sense to obtain patent protection for a device which he intended to produce. In fact, his attorney is said to have urged him to generalize his application to cover all substances with useful sensitivity. In actuality, the applications went in one at a time over a period of months, each covering a specific case. As for Dunwoody, he and de Forest needed a detector immediately, or they would have been forced to halt receiver production after losing the patent lawsuit against them concerning their electrolytic detector. There is no doubt that they filed to patent the carborundum detector as soon as possible.

Ferdinand Braun was also directly motivated, even though he was an ocean away from American developments. He was now associated with the Telefunken organization, the latter being the result of merging his company and others by directive of the German government. The object of the consolidation was to improve the German competitive position relative to foreign competition, especially the British and American Marconi.

Telefunken management realized that better equipment was necessary if they were to remain competitive. In the case of receivers, it was decided to develop Braun's crystal detector, even though it meant the end of previous commitments. Two other considerations may have contributed to this decision, even though they were based on rumors later found untrue. The first of these was concern about another German invention.

A young researcher named Robert von Lieben was working upon a three element tube, predating de Forest's invention of the triode. Von Lieben's procedure and goal were somewhat different, because he was interested in amplification rather than detection. Nevertheless, Telefunken management perceived von Lieben's work as a possible threat against their current detector technology, and they wanted to counter it. (Telefunken and the von Lieben interests later cooperated in development of the tube.)

A second concern stemmed from ongoing press commentary about the work of Nikola Tesla in the United States. Tesla was somewhat later relegated to the rank of visionary, but at that time it was still fresh in mind that he had single-handedly revolutionized the power industry by inventing the polyphase AC generation and distribution system. Tesla was never shy about promoting his ideas and had conducted spectacular demonstrations involving high frequency power and wireless transmission.

As early as 1902, Tesla had directed construction of a huge tower near the town of Shoreham, Long Island, and he had announced a forthcoming system, a "World System" of multiple uses of wireless power transmission, including transoceanic broadcasting. Tesla was encountering serious financial difficulties by 1906, but until just a few months earlier there had been an almost continuous stream of favorable publicity, marred only occasionally by bizarre speculation. It has been suggested that Tesla's concept of a World System, which included a "World Wireless System" as part of the package, was also a factor in Telefunken's decisions. Tesla did, after all, show a major construction investment on Long Island, and there were eyewitness reports of tests in progress that demanded very large amounts of power.

Whatever the reasons, Telefunken management gave Braun the authority to proceed, and he chose to patent a detector using psilomelane, a fairly common mineral composed mainly of manganese dioxide. It is not clear whether his choice was based upon locally available mineral deposits with exceptionally good properties, or whether his specimens of other minerals were of poor quality. In any event, the mineral he selected never came into use in the United States. It's post-patent history in Germany is not known at present. It is known that Telefunken later used carborundum crystals for at least some of its detectors, after reaching cross licensing arrangements with the patent holders.

The applications for patents and immediate practical usage made 1906 the year that the crystal detector arrived as a commercial receiver component, and this, of course, made the crystal set a reality. The latter soon became the most important type of receiver in use, and it remained so for several years. This was true in spite of the fact that the Marconi Company in England was already using a vacuum tube (the Fleming Diode Valve) and that de Forest had already applied for a patent on a similar design, resulting in more litigation.

Further, it was as early as January 28, 1907 that de Forest applied for his triode Audion patent. It was only later recognized for what it really was, the forerunner of the high vacuum triode that radically changed the entire communications industry. Curiously, even de Forest did not seem aware of its importance for application as a detector until about 1910-1911, and so the crystal detector dominated the scene until World War I. It was still used throughout that conflict, concurrently with other alternatives. It was not unusual to see a manufactured receiver complete with one or more crystal detectors, plus a panel mounted vacuum tube and suitable provisions to switch from one to another.

Continuing Development

The first patents did not signal an end to crystal detector development. Pickard continued to investigate methods and materials for improvement. He had chosen the combination of a pointed brass wire pressing against silicon, which he called a Perikon detector (Perfect Pickard Contact). Later the term came to be more closely associated with another Pickard invention: a blunt pointed crystal, held in contact with the surface of a second crystal of different composition. He procured a registered trade name for the term, but the name is now generally used for any combination of two crystals in a Perikon type arrangement, regardless of origin. Pickard patented the Perikon combination of zincite and chalcopyrites, which was used to some degree along with the other detectors.

The de Forest Company also continued work on the crystal detector. The first method of making connections used by Dunwoody was soon replaced by a more practical design by an employee of de Forest, C. D. Babcock. The new device was a holder which clamped a crystal fragment between two flat metal surfaces with a provision for changing the clamping pressure by means of an adjustment screw. This arrangement also permitted changing contact locations by loosening the screw and moving the crystal. Babcock applied for a patent on August 27, 1906, only five months after Dunwoody's application date, so it is evident that the de Forest interests were continuing their development effort.

It is also evident that de Forest was still in trouble. Babcock's design provided the equivalent of a point contact by relying upon the jagged surface of the crystal fragment touching the metal surface of the holder. A problem arose from the fact that this was true of both connections rather than just one, and for mechanical stability as many as three points at each connection would usually exist. In short, the de Forest investigators were still unaware of the best mounting method for a crystal: a point contact at one connection and a wide area surface contact at the other.

The magnitude of the problem became all too obvious when the de Forest Company attempted to produce the carborundum detector for commercial applications. Only about one in ten of the production units performed satisfactorily, and nobody understood why. The solution was quite appropriate to the occasion: Pickard was retained as a consultant to solve the dilemma.

There is some question as to just when Pickard himself had recognized the importance of a wide area electrical contact at the base of the crystal and a point contact at the other, but there is no doubt about his success with the carborundum detector. His new design consisted of a stiff pointed contact, held under pressure against the active surface of a carborundum crystal mounted in a button of solder. This provided the optimum contact condition at both connections. With these changes, the carborundum detector was produced with a much lower rejection rate. He also introduced the bias battery as a method of considerably improving detector sensitivity, and he later received a patent on his innovations. This improved detector was used for several years by the de Forest Company, and by its successor, The United Wireless Telegraph Co.

Pickard and his attorney, Phillip Farnsworth, started a small company to manufacture and sell crystal detectors. In February of 1908, they, together with a retired army colonel named John Firth, incorporated the company as the Wireless Specialty Apparatus Company. They produced crystal detectors and other components at first, and later expanded to manufacturing complete radio receivers. Their detectors were tailored for marine applications, which at first constituted their principle sales market, although the U.S. Army Signal Corps was also a customer. Their equipment was of the highest quality and is much sought after by present day collectors.

There was another outlet for sales which began to assume some importance during the first decade of this century. A growing number of amateurs in the United States were experimenting with wireless. Most of these were boys and young men with limited funds, and who often built their equipment from whatever was at hand. One of the few things they could afford to buy outright was a good crystal, which was an inexpensive as well as sensitive detector. As they matured and had more money, they began buying detector stands and other manufactured components. This amateur market appears to have developed very early after the introduction of the crystal detector in late 1906. As early as 1907, one of the pioneers of radio, Lloyd Espenshied, himself an early amateur, was already using a crystal detector.

During this period, at least one, and very likely a number of experimenters, discovered that using a fine wire catwhisker, rather than a heavier pointed wire, improved the sensitivity of a number of different detector materials. In spite of that improvement, nobody filed a patent application for the catwhisker contact until Pickard took advantage of the situation and filed in 1911. Years later, Pickard made the following comments concerning the invention of the catwhisker: "I have no idea who first used a fine wire contact with materials requiring low pressure contact, and by 1911 my company had produced and sold a number of fine wire contact galena detectors to the U.S. Navy, and in the same year I filed a patent application for the 'catwhisker' detector which issued later as U.S. Patent 1,104,073. I have no doubt that somewhere around 1911, or perhaps even earlier, the same idea occurred independently to others." The catwhisker had one big disadvantage. Anyone who has wrestled with the adjustment of a catwhisker, first searching for a sensitive spot on the crystal, then maintaining good contact, will understand why the catwhisker was never very popular onboard ship. It is very sensitive to vibration and shock and easily knocked out of adjustment. Marine operators faced a formidable problem working with the device, especially in rough seas.

Some of the minerals used for detectors were subject to still another drawback, which the use of a catwhisker tended to magnify. Heavy static or operation near a transmitter sometimes destroyed the sensitivity of certain materials at the point of contact. Some onshore stations were built with the receivers located at a distance from the transmitter, at least in part because of this problem, but this could not be done on a shipboard installation. Of course, these difficulties did not prevent the growing demand in the amateur sales market, and for other limited applications.

The Competitive Years, and the Role of the Crystal Detector

The small American Companies, which struggled through the first few years after the turn of the century, either failed, changed hands, or consolidated, so that the business scene had radically changed by 1910. One of these changes was the acquisition of the de Forest Company by a new company, The United Wireless Telegraph Company, followed by rapid growth of the latter. Most of the de Forest assets went to United Wireless, including Dunwoody's carborundum detector. In the brief period between 1907 and 1911, United Wireless grew to the point where it was the dominant company in the United States, having far outstripped it's principal rival, the American Marconi Company, a subsidiary of British Marconi.

The Marconi interests were producing receiving equipment which depended upon the Fleming Valve, a vacuum tube diode which performed better than their previous detectors. Unfortunately for Marconi, it had become evident that the Fleming diode was not as satisfactory as the crystal detector, and the same problem was also evident in comparing much of the other Marconi equipment with the competition. However, the sheer size of British Marconi and its subsidiaries meant that the Fleming Valve was more important worldwide than the crystal detector. All of this was soon to change, because of an important event that took place in 1912.

At that time, American Marconi filed suit against United Wireless for using certain Marconi patents without permission or licensing agreements. There was considerable justification for doing this, and United Wireless was well aware that they would lose the resulting court decision. Therefore, they did the only sensible thing they could do under the circumstances, which was to plead no contest and try to reach an out of court agreement with American Marconi. An agreement did result, in which American Marconi acquired virtually all of the assets of its much larger rival, and suddenly had the advantage of a near monopoly in the United States for the manufacture and sale of wireless equipment.

One consequence of this dramatic turn of events was that the carborundum detector became available not only to American Marconi but to the even larger parent corporation in England. In fact, the Fleming Valve was soon replaced by the carborundum detector, and the latter became the standard detector throughout the Marconi business empire for several years.

The Wireless Specialties Company was a small independent corporation not directly affected by this situation, and it was as its name stated, strictly a specialty company. It continued to supply the crystal detector types protected by Pickard patents, and to build and sell high quality receiving equipment. There was no effort to penetrate the high power transmitter segment of the wireless business. In those days, manufacture and construction of large transmitter installations was analogous to the "heavy industry" portion of business in general, due to the huge size, power requirements, and complexity of supporting facilities.

American Marconi continued using the carborundum detector until 1913, at which time it began offering a receiver of improved design employing both a carborundum and a cerussite (PBCO) detector as alternative choices. The cerussite crystal was claimed to be as sensitive as a galena detector, which is now considered to be the most sensitive available substance. Most of the well known detector materials were in use by 1910-1913, although it would be many years before germanium and high purity, precisely doped silicon would be available. Pickard's and Dunwoody's inventions both survived and were commercially successful.

The contributions of Braun and Austin were never utilized in the United States, and Bose's work had not affected the course of events. The understandable but misleading tale that Pickard and Dunwoody were the sole investigators in the field probably stems from that situation.

The major part played by the crystal detector just before World War I resulted from several conditions. First, the Fleming Valve was not as sensitive as a good crystal. This was one reason why the Marconi companies took the opportunity to use carborundum after the acquisition of the United Wireless assets. Further, the crystal detector was easy to produce, much less fragile than a glass vacuum tube, and did not require considerable power to heat a filament. The de Forest Audion, first as a diode similar to the Fleming Valve, and shortly thereafter as a triode, had been invented almost concurrently with the carborundum detector, but the triode version had serious problems. It was not, in fact, recognized at first as a practical device. This was in spite of considerable publicity and promotional efforts. De Forest himself delayed several years before proceeding further with it, and even then found it to be unreliable. Audions were never mass-produced with closely similar characteristics, and the characteristics of any one tube sometimes changed unexpectedly during use. Competition between the crystal detector and the triode was not important until several companies in the United States and abroad realized the possibility of developing a high vacuum triode for an amplifier tube as well as for simple detection. This did not occur until about 1912.

Other devices had been proposed as detectors, and some were patented and used in varying degrees. Marconi's magnetic detector was reliable, but it became obsolete because it was too insensitive. The electrolytic detector was successful using liquid electrolyte, but the latter, usually nitric acid, spilled easily with unpleasant results. The solid electrolyte version did not have the spillage problem, but it remained in obscurity for unknown reasons. It can be speculated that it was too insensitive, as was mentioned earlier. The heterodyne detector, proposed in 1902, showed great promise, but its application was delayed during the 1902-1910 period because of lack of satisfactory operating components. An electromechanical system called the tikker (or ticker) did reach a degree of limited use. A number of other devices were patented, but they were found to be impractical.

World War I brought with it a transition period with respect to the crystal set and other receivers. An unusual state of coexistence was evident among those in use at the outset of the war. These remained in use concurrently, and in the case of crystals and tubes, both were used in new receiver designs as the war progressed. Improvements in vacuum tube design, and the invention of circuitry which took place just before and during the course of the conflict, later changed this state of affairs. Specifically, the invention of the regenerative detector, actually a receiver circuit using the improved high vacuum triode, was very likely the principle reason for this change. Such receivers showed dramatic improvement in performance compared with the crystal set.

By the end of World War I, the superiority of the regenerative receiver, and also the advantages of amplification using the new tubes, were widely recognized. This new technology was hampered at first by the legal restrictions of patent law and a storm of litigation, but this did not prevent the obsolescence of the crystal set. Its days were numbered as far as the "traditional" business of point-to-point communications was concerned. In that particular field, especially for shipboard installations, the crystal set assumed the function of an emergency back-up unit, available when all else failed. This was a minor application, but a small demand did exist for many years. It was a post-World War I phenomenon of a different nature that extended the life of the crystal set as an important commodity.

A New Industry

The amateur market grew rapidly after the war, providing a sales outlet for crystals, but what really rescued the crystal set from obscurity for a few more years was broadcast radio. 1920 marked the beginning of the growth phase for this new industry and of its importance as part of the American scene. Within just one decade, broadcast radio created an enormous market for home receivers and affected the habits and customs of the general population.

Crystal sets were very popular at first, because they were inexpensive and easy to operate compared with more elaborate radios. For example, in 1922 manufactured crystal sets were available for $25 and up, whereas the least expensive one tube receiver sold for about $75. Crystals were also used in some sets together with tubes. Most such usage occurred during a brief period in the early 1920's when interest in radio assumed some of the characteristics of a fad or craze. At times the demand for radios outstripped the supply. Unlike a fad, however, home radio sales matured into a permanent component of the total consumer market. This created a major manufacturing and sales industry, which supplied a public that became more and more sophisticated in its preferences as time went by. This became an important factor which affected the evolution of the radio.

Most broadcast band receivers built before 1923 were constructed at home, and ranged from simple crystal sets to one to three tube designs. Inexperienced family members often found them difficult to operate, and the sets using tubes often required bulky and expensive batteries. Manufactured radios were a small fraction of the market at first, in spite of the great demand, because it took time to organize a new company, and then design, test, and set up a production line. Many new companies were started, but it was not until 1925 that annual commercial production equaled the number of home built radios during the year.

Consumer demand favored the latest improvements in convenience and performance, and this, plus rising competition, pushed a rapid development of better receivers. That trend doomed the crystal set as a major market seller. The crystal set was acceptable while there was just one station in town, but as more and more stations came on the air, as multiple frequencies were assigned, and when a fad for long distance reception blossomed, the inferiority of the crystal set became obvious. It was too insensitive, selectivity was poor, and listening with earphones was awkward for large families. The multi-tube radio was the big winner here, even though it meant the end of the era when most receivers were home assembled.

After 1921, the manufactured radio dominated the retail sales market, and its design and production had become an important engineering specialty which continued to evolve during the 1925-1930 period. The biggest single improvement was the introduction of AC operated radios which performed well, and at the same time could be purchased at a reasonable price by the average family. That milestone was reached in 1928, and by that time the crystal set was a thing of the past, as was the crystal detector itself in most cases. In fact, by 1930, products using crystal rectifiers for any purpose were a very minor percentage of the total market for all types of radio equipment. A minor market did exist, nevertheless, and deserves discussion. Beginners in radio, as well as hobbyists, created a small but stable sales outlet for many years. There was also activity in the sales of a miniature type of radio best described as a toy or novelty. These topics are covered in the next Section.

Survival as a Specialty

Crystal set circuitry was an attractive method to introduce students to electronics for many years, and the diode detector circuit is still in the curriculum. The detector stand, crystal, and catwhisker have long since been replaced by a germanium diode, but the basics are still the same. What is important to the suppliers is that the market may be small, but at least it is stable. Barring utter catastrophe, each year sees a new class of beginners, even though the numbers may fluctuate with demographic trends. In this situation a solid-state diode and associated circuitry continues to play a very minor but very long duration teaching role for now and the future.

Another small but more dynamic market has been the supply of crystal set related hardware and information to hobbyists and collectors. It is reasonable to relate the origin of such business to activity of the early 1920's, when the crystal set was popular as a home radio. Many were purchased part by part and assembled at home, so it was profitable for suppliers to stock appropriate materials and components. As radios were improved, the home-brew crystal set market shrank, but there were still numerous sources for parts through the

1930's, and even after World War II. The local radio repair shops had new stock, and there was an abundance of used and discarded radios, available at nominal cost, which could be sold for parts.

There was another important source which has since disappeared, along with the local radio shop which is now all television. This source consisted of a number of national and regional mail order suppliers, and for over four decades they offered just about anything that might be needed. The national suppliers included such names as Allied Radio Corporation, Bernstein-Appleby, Concord, Lafayette, and Radio Shack. Most of the larger suppliers remained active until the 1970's or '80's, but they eventually went out of business or changed their product lines to concentrate on complete receivers, high fidelity equipment, industrial electronics, and even non-electronic appliances and equipment.

The demise of the mail order components business was not due to the decline of the crystal set, which had occurred many years earlier. Even before 1930, most parts sales were to do-it-yourselfers working on more elaborate equipment. There was even a modest business in television parts and kits for a time after World War II. There has been some speculation concerning what caused the decline in parts sales. One hypothesis suggests that the rise of solid-state technology resulted in such compact, highly efficient, and professional looking products, that the home brew enthusiasts felt it impossible to match or exceed the commercial equipment. At present there is an intense interest in electronics, but often the interest lies in assembling a complete system, using building blocks such as tuners, amplifiers, playback units, and so forth, with no curiosity as to what goes on within the individual blocks. Therefore, "black boxes" still sell, but not the parts inside.

Whatever the reason, there was a period of years during which crystal set components were scarce. Hobbyists should thank the growing collector interest in early day radios for improving the parts situation, even while they are removing older radios from availability. Collecting, which includes restoration, reproduction, and replicating, became popular enough during the 1980 decade to encourage small businesses to begin supplying parts and information again. At first, supplies were from "new-old-stock" or were used. However, now in the 1990's, there is a limited production of components and small hardware needed for crystal sets.

One small mail order business specializing in the crystal set field began during the unlikely depression period in about 1932, and persisted through the lean years that followed. This was Modern Radio Laboratories (MRL) now located in Minnesota, although originally a California based company. MRL was founded by Elmer G. Osterhoudt, a 1920's era wireless operator, amateur, and later a radio serviceman.

After establishing MRL as virtually a one man operation, Osterhoudt became known as an expert in practical knowledge of the crystal set field. He published and copyrighted much literature on the subject, all tailored for both novice and experimenter, which was sold through his company. This included pamphlets, a magazine-type periodical, catalogs, and other information. The catalogs offered a wide range of components to his customers, which were ideal for crystal set applications. Also sold were kits of his own design.

MRL survived the years when larger companies left the mail order business, and even produced its own simplified detector stands and switch point levers when these became otherwise unavailable. Selected components, which could be substituted for hard-to-find parts needed for new construction, were also located and featured. Osterhoudt's death following a 1987 automobile accident caused the company to suspend its operations for a time, but, about a year later it was sold and moved to Minneapolis to resume operations.

Another, rather different type of business lingered on for many years after a strong beginning in the 1920's. Small crystal sets were cheap and sold well, and some of these were reduced to the minimum possible circuitry. The dominant company in the field was the Philmore Manufacturing Company, which had started earlier as the Ajax Products Company before its incorporation in 1925. Their cheaper products were small open type sets, with outside-mounted detector stands or holders for the crystal. These sets were of a size and design that actually placed them close to a novelty and toy category, and later their entire product line veered toward that type. They did continue to sell separate components, chiefly a detector stand, which was still available as late as 1956.

There were other companies in the field at one time or another during the long span of Philmore's activity, some of them actually doing business before Philmore started. None seems to have lasted as long, however. There was also a true novelty and toy market in the 20's, supplied by several small businesses. Another wholesale outlet for the manufacturers was the supply of premiums advertised to boys as an inducement to sell a product or eat a certain brand of cereal.

Good quality crystal sets disappeared from the stores as the public demanded better performance, but the novelty and toy market continued for over thirty years. In fact, it may not have reached its peak until after World War II, when advertisements promoted tiny crystal sets as a "new technology." Except for that emphasis, the type of advertisement remained almost the same during the entire era of novelty manufacturing. Even in the early 20's, advertisements picture a boy grinning with excitement as he listens to his tiny set, also prominently pictured, and the same general format was used for years. Early promotions also emphasized some of the set's technical details, but later it was simply stressed that no batteries were ever needed. The low price was also featured, and for premiums the "free" inducement appeared in big bold letters. The fact that an external antenna and ground were required was either not mentioned at all or downplayed.

The Philmore company was in the novelty and toy market from the early days, but a later rival appears to have out produced it in more recent years. Western Manufacturing Company of Kearney, Nebraska made these radios using a number of brand names and small company names for firms that they either owned outright or controlled. Some other companies also participated, although they appear to have been less prolific.

There were manufacturers that took advantage of what was probably the most famous miniature radio of all time, the Dick Tracy comic strip wristwatch radio. The latter was a fictional two way communications set, but Da-Myco Products Company of New York City made a wristwatch crystal set and advertised it as the Dick Tracy Wrist Radio. Da-Myco advertised the receiver as a toy, but there were some competitors who were not quite so truthful.

More generally speaking, the appearance of most novelty and toy radios evolved from the early day arrangement that usually had an exposed crystal detector stand or holder, to a modern looking little box, often made of colored plastic, and shaped either like a miniature table model radio, or to some novelty form. This evolution became practical for even the cheapest sets with the advent of the solid-state diode, described in the next Section.

Mass-Produced Rectifiers

The decline of the crystal detector in its original form did not mean the end of the crystal rectifier. Two important developments in electronics saw to that. First came laboratory research in the 1930's which led to radar for military applications. During that time it became evident that the crystal rectifier would be the only satisfactory solution for certain applications at microwave frequencies.

After initial work in the laboratory with hand-made rectifiers, a concerted effort was made to develop a miniature fixed diode package that could be mass-produced for operational radar sets. The first such diodes used silicon. Later on, diodes were developed using germanium, the semiconductor element not available in the early days of radio. The engineering required to fabricate packaged diodes was a considerable achievement in itself, and the research and development effort to produce silicon and germanium with the required properties was even more impressive.

The minerals and other substances for early day crystals were used in their natural state, or as produced, and the best performing crystals were chosen for service. Selection was often made by the radio operator, searching for sensitive spots on each crystal. Advertisements for mounted crystals sometimes specified them to be pretested, but this could mean about anything in those days. Either method would be impractical with modern diodes.

The substances for modern applications had to be highly purified and then doped with very small but precise amounts of one or more other elements, in order to meet rigid quality control standards. Mass-production without such standards would result in unacceptably high rejection rates. The time and expense needed to develop acceptable silicon and germanium crystals probably exceeded all of the previous efforts that resulted in the original crystal detectors.

By the time that World War II ended in 1945, the production capability was at hand to provide crystal diodes for civilian applications. This meant that the germanium diode was conveniently available when the second major development in electronics that would affect its use took place. This, of course, was the invention of the transistor, by William Shockley, John Bardeen, and Walter H. Brattain, all at the Bell Laboratories subsidiary of American Telephone and Telegraph. The invention was first demonstrated on December 23, 1947, and it was considered so significant that the trio was jointly rewarded the Nobel Prize for Physics in 1956.

Like so many inventions, the transistor did not immediately revolutionize an industry, electronics in this case. It took lengthy efforts to improve the transistor and to develop mass-production methods. It was late 1951 before the manufacturing subsidiary of AT&T began full-scale production, joined by the Raytheon Corporation a little later.

In fact, it was several years before the transistor became a major competitor with the tube. During that interim period in the 1950 decade, sales of crystal diodes developed to some degree on their own for various requirements.

The first transistorized radios were portables, where the advantage over tubes was highly evident. These were first produced in the 1950's, and were not the sudden and overwhelming success that some had envisioned. The general public was not much interested in portables at that time, so only modest sales were realized. The trend was there, nevertheless, and within a few more years fully transistorized radios and television sets were the rule rather than the exception. This was the great opportunity for widespread use of the germanium diode for detector as well as other applications, since it made no sense to retain the vacuum tube for that one purpose. The silicon diode also made a reappearance, for uses where its properties were more suitable than those of germanium.

Since the 1950's, literally hundreds of different solid-state diodes have been developed for as many specialty applications, and diode detector usage continues. The even more recent development of integrated circuitry has added another physical form and still smaller size to the array of diode configurations, but the same basic principle still applies. In the long run, it has turned out that the nineteenth century discovery of unilateral conductivity did not remain the scientific curiosity that became the fate of some other discoveries of the time, but led to the broad applications of the present day.

BIBLIOGRAPHY

Aitken, Hugh G. J., Syntony and Spark, the Origins of Radio. Princeton University Press, Princeton, New Jersey, 1985.

Barnouw, Erik, A Tower in Babel. Oxford University Press, New York, 1966.

Blake, George G., History of Radio Telegraphy and Telephony. Chapman and Hall, London, 1928.

Briggs, Thomas H., The Triode that Pre-dated de Forest: Robert von Lieben and the LRS Relay. The AWA Review, Vol. 5, 1990, p. 45.

Cheny, Margaret, Tesla, Man out of Time. Dorset Press, New York, 1989.

Constable, Anthony, Early Wireless. Midas Books, 12 Dene Way, Speldhurst, Tunbridge Wells, Kent, England, 1980.

The Continuous Wave, Technology and American Radio, 1900 - 1932. Princeton University Press, Princeton, New Jersey, 1985.

Denk, William E., Jagadis Chunder Bose and his Galena Detector. The Old Timer's Bulletin, Vol. 27, No. 1, June, 1986, p. 16.

Douglas, Alan, The Crystal Detector. IEEE Spectrum, April, 1981, p. 64.

Douglas, Susan J., Inventing American Broadcasting, 1899 - 1922. The Johns Hopkins University Press, Baltimore, 1987.

Dunlap, Orrin E., Jr., Radio's 100 Men of Science. Harper and Brothers, 1944. Reprinted by Books for Libraries Press, Freeport, New York, 1970.

Espenshied, Lloyd, 1907 Amateur Station. The Old Timer's Bulletin, Vol. 31, No. 1, May, 1990, p. 11.

Kurylo, Friedrich, and Charles Susskind, Ferdinand Braun. The MIT Press, Cambridge, Mass., 1981.

Lewis, Thomas S. W., Empire of the Air. Harper Collins Publishers, New York, 1991.

Maclaurin, W. Rupert, Invention and Innovation in the Radio Industry. MacMillan Co., New York, 1949.

Mayes, Thorn L., Wireless Communication in the United States. The New England Wireless and Steam Museum, Inc., East Greenwich, Rhode Island, 1989.

McNicol, Donald, Radio's Conquest of Space. Murray Hill Books, Inc., 1946. Reprinted by Arno Press, New York, 1974.

Pickard, Greenleaf Whittier, How I Invented the Crystal Detector. Radio Age, Vol. 2, No. 10, Dec. 1976, p. 1. Reprinted from Electrical Experimenter, Aug., 1919.

Schiffer, Michael Brian, The Portable Radio in American Life. The University of Arizona Press, Tucson, Arizona, 1991.

Sievers, Maurice L., Crystal Clear. The Vestal Press, Ltd., Vestal, New York, 1991.

Thackeray, Desmond P. C., When Tubes Beat Crystals: Early Radio Detectors. IEEE Spectrum, March, 1983, p. 64.

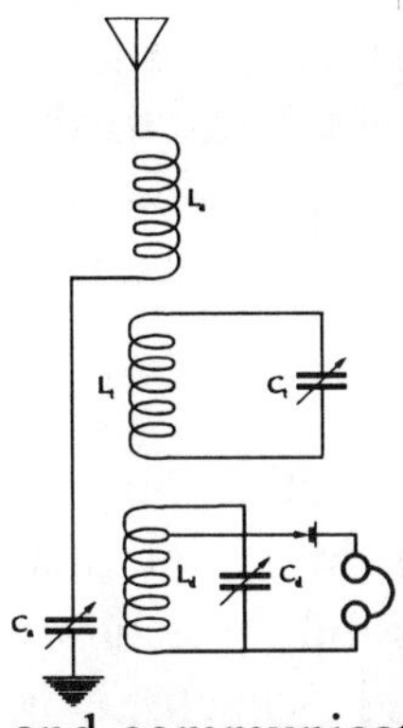

CHAPTER 2

FUNDAMENTALS OF RECEIVER SYSTEMS

Some Comments on Definitions

The crystal set certainly played its part in the history of radio, as told in chapter 1. Yet it lives on in each and every receiver and communication system that exists today. Crystal detectors and the crystal set as a whole laid the foundation on which all receiver systems are based. Hence, in order to design these systems, or to design an advanced crystal set today, these fundamentals are still the first principles that you must understand. That is how this chapter fits in; it is the transition between the idea of the crystal set (and its history) and designing and/or building sets. This is the theory class. The lab, or design chapter, follows.

The terms "crystal set" and "crystal detector" originated at the time when detector materials were often naturally occurring crystalline substances. Detectors were not always natural substances however; amorphous, semimetallic, and fused solids were sometimes used. The term "mineral detector" has been employed, but the same exceptions still apply. In short, crystal, mineral, and other words that have been suggested are all misnomers. Nevertheless, the word crystal is so firmly attached to the subject that it will continue to be used here in spite of its shortcomings.

There has been some lack of agreement in defining what a crystal set actually is. It is defined here as a receiver employing a crystal detector, which operates only on signal power delivered by the antenna, the only exception being for those detectors which require bias voltage at the crystal to improve their sensitivity. The exception dates back to the earliest days of the crystal detector, when it was found that bias voltage on a carborundum detector improved its performance.

There are reasons for the narrow limits on the definition of a crystal set. One of the most obvious comes from the fact that no one thinks of a transistor radio using a solid state diode detector as a crystal set, so it is not considered as such. But, consider a less clear-cut example. Is a crystal set still a crystal set when an audio amplifier is connected at its output terminals? As defined here, the answer is still no. Amplification with the help of external power would improve the performance of almost any crystal set, but for some the excitement of distant reception by signal power alone would be missing. Further, if improved performance is the goal and amplification is acceptable, why stop at adding audio to an existing crystal set? Why not optimize the detector-amplifier combination as a system? The design challenge is there; the resulting improvement would be quite evident. It is simply a different category of challenge.

On the other hand, suppose that an active element such as a transistor is present, but it is not externally powered. Here, this is still a crystal set. The system is limited to power supplied by the antenna; the achievement of distant reception without external power is still possible. Although no one has demonstrated that an active element can be used to advantage this way, there are some possibilities that can be investigated.

A self-powered transistor circuit with crystal detection has been suggested. One such arrangement consists of two crystal sets in one: one tuned to a local station to develop a supply voltage and the other as the crystal set that includes a transistor audio amplifier. The voltage obtained by rectification of the local station's signal is used to "power" the transistor of the crystal set. The sets would share the same antenna. Still other possibilities are the use of an active element as an impedance matching device, or as a means of improved coupling between a tuned circuit and the detector. These are all crystal sets as long as the ban on external power is observed.

Concerning other definitions used here, most of the terminology of modern electronics applies, and the reader with a background in basic electronics will usually feel at home. There are a few instances, however, where component names and units of measurement have changed over the years. Also, some components and circuits are used in crystal sets which are rarely if ever found in present day equipment. These components will be defined as needed.

Signal Reception Requirements

No specialized knowledge is necessary to understand that a device which is completely dependent on transmitted power needs to capture as much of that power as possible for best operation. In general, this need translates into operating with a long antenna and a good ground connection. These are requirements that differ dramatically from what is satisfactory for today's typical receiver. The antenna and ground are so vital to the operation of a crystal set that they are treated here as parts of the overall system.

The basic crystal set antenna is the inverted L, consisting of a long, straight, horizontal length of wire which picks up the transmitted radio frequency signal, and a short vertical length, which is the lead-in. In early day terminology, the horizontal wire was called the aerial, and the aerial plus lead-in combination was called the antenna. Modern wording is used here.

Antennas

The inverted L antenna is somewhat directional, being most sensitive to signals arriving parallel to and from the distant end of the antenna. See Fig. 2-1. As a general rule, longer is better as far as the horizontal length is concerned, with the lead-in kept as short as possible. The minimum length that will pick up a signal other than from a local station is usually some 40 to 50 feet, and a decided improvement results if it can be extended to 100 feet or more.

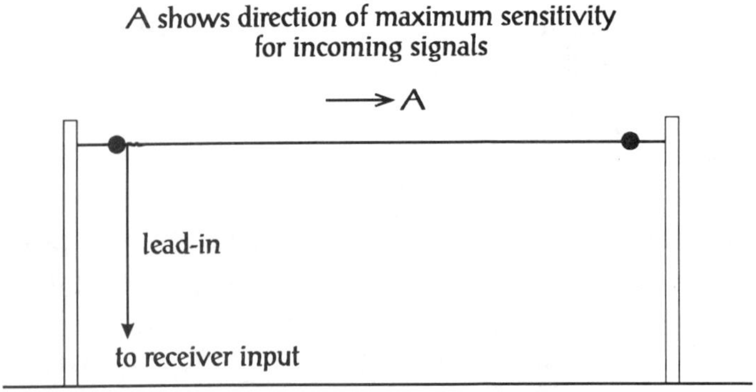

Figure 2-1: Diagram of an inverted L antenna and lead-in, giving the direction of maximum sensitivity.

Ideally, the antenna should be 20 to 30 or more feet high, and kept well away from buildings, trees, power lines, and any other large improvements on the land. However, this is not always practical except for the lucky few that live on a large open tract of ground. Usually a compromise layout must be made for a city lot, or where other buildings, hills, or trees dominate the landscape, and this compromise extends to both length and height. Fortunately, there are some alternatives that help the situation. Sometimes a long, low antenna works better than a short, high one. Changing the lead-in end or the antenna direction can make a big difference. Of course, all of these options depend upon antenna site limitations.

Another type of antenna with some of the characteristics of the inverted L is the vertical antenna. Its big advantage is that it can be erected where there is practically no room for anything else. Its big disadvantage is that some kind of mast is needed to hold it, unless there is an appropriate existing structure available. Such an antenna must be at least 30 to 40 feet in vertical length to have characteristics comparable with the shortest usable inverted L. A wire running down the side of a building of this height and a few feet from the wall is one possibility. Dropping the wire from a heavy horizontal line between two buildings is another. A mast of adequate height is always a solution, but it does require considerable construction unless a kit is purchased. These are available from suppliers that specialize in a full line of amateur radio equipment, because they are used for transmitter antenna masts.

The chief difference between a vertical and an inverted L antenna is that the vertical antenna is omnidirectional. That is, it is equally sensitive to signals from all points of the compass. This may or may not be an advantage, depending upon individual requirements. The lead-in can be very short if the lower end of the vertical is not too high and if it can be located close to the receiver. The unique vertical layout requirements often make this possible.

The antennas described so far have the lead-in attached at one end. The name "end-fed antenna" is sometimes applied to this category, although the term is more appropriate for transmitter antennas. End attachment is not a necessity, however, because the lead-in can also be placed at the midpoint. The simplest arrangement of this type is the T antenna.

A T antenna, shown in Figure 2-2, is bi-directional, being equally sensitive to signals from either direction along the wire. It is a very advantageous type for problem installations requiring receiver location below the antenna, rather than near or beyond one end. This shortens the lead-in, which can make a considerable difference in performance.

An L or vertical antenna cut to one-quarter wavelength (for the frequency of reception) will be resonant. In fact all antennas cut to one-quarter will be resonant. The advantage of doing this, if space allows, is that the antenna can then be tuned with the receiver for the reception of shorter wavelengths (higher frequencies) too. Hence, the ideal L or vertical would be cut to one-quarter wavelength for the lowest listening frequency desired. For example, by cutting the antenna for the bottom of the AM band, it is possible to tune it throughout the band. The T antenna, on the other hand, would have to be doubled in length; each side must be one-quarter wavelenth long. This is a serious disadvantage if the T is to be used for the broadcast band - and you don't have much space!

Another type of antenna that is very similar to the T in many respects is the dipole. Instead of a single connection at the mid-length, the two halves of the total length are separated by an insulator, and the output of each half is fed separately to the receiver. This type has been seldom used with crystal sets, although it is highly favored for other applications.

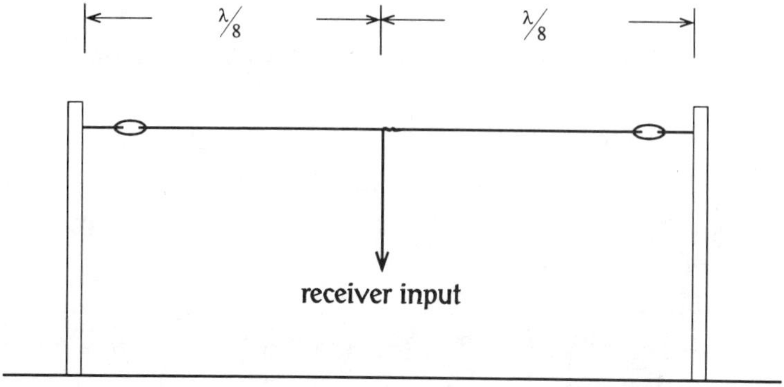

Figure 2-2: A typical T antenna.

All of the antennas previously discussed are broadcast band tunable at the receiver, provided that their length is something like 40 to 50 feet or more. *Antenna tuning* is a *highly desirable feature* that will be discussed later in detail. If you are interested in a highly selective set that still retains reasonable sensitivity, it is a definite advantage to have it. Nevertheless, circumstances sometimes prohibit an outside antenna or even a long single wire in the attic or down a hallway. There is one other tunable type that can be used in a room, and that is the loop antenna.

In situations where a straight wire antenna cannot be used, then a nonresonant antenna can be considered as a last resort. The advantage of antenna tuning is lost, but it does work in most cases, sometimes all too well if there is a strong local station. A broad band of interference with other stations results. A nonresonant antenna consists of one or more wires strung around the walls of a room, the ceiling, or in an attic, with bends and corners as necessary to increase the total length of wire. Even electrical wiring can be used, but remember to install a series blocking capacitor and current limiting resistor for safety.

All types of outside antennas share common requirements, in terms of construction, in order to withstand continuous exposure to the elements while functioning properly at the same time. Practical problems of choosing the proper wire, use of insulators, and installation technique are discussed in this and a following section.

Antenna wire should be of large diameter, both for good electrical conductivity and for strength. The best choice for lengths up to a hundred feet or so is No. 7-22 insulated copper. This consists of seven twisted strands of No. 22 American Wire gauge (AWG) copper wire with overall rubberized or plastic insulation. The insulation serves as protection from the outside atmosphere. Bare wire can be used, but it has a lifetime that depends upon the environment. In the dry southwest it may last for years in rural areas. Under harsh marine air conditions its life may be measured in months.

The stranded wire recommended here has good strength characteristics to withstand wind and icing conditions. Solid copper can be substituted, but large diameter wire must be used to provide the same strength as the equivalent stranded type. No. 14 AWG single conductor copper may be used for the 100 foot category, and No. 12 AWG for 200 feet. The breaking strengths are 213.5 and 337.0 lbs., respectively. A copper coated steel wire, trade named Copperweld, can be used for very long lengths of several hundred feet. Its effective conductivity is near that of solid copper while it has a much greater tensile strength. No. 14 AWG Copperweld, with 30% copper, has a breaking strength of 440 lbs., over twice that of the equivalent 14 AWG solid copper.

Copperweld and similar copper clad products conduct high frequency alternating current almost as well as solid copper. This is due to a phenomenon called skin effect, where current at radio frequencies is high at or near the surface of a conductor and decreases toward the center. Skin effect permits the substitution of copper clad steel for pure copper without significant losses at radio frequencies. Copperweld is available in a range of sizes for both stranded and solid core wire.

Stranded wire is less susceptible to damage during shipping and handling than solid core, and heavy insulation decreases the likelihood of damage even further. All wire should be inspected for defects before installation, and this is especially important for solid core bare or enameled wire. That type is particularly susceptible to kinks. A kink seriously cuts the strength of wire, for both breaking strength and fatigue. Fatigue strength is the resistance to repeated application and removal of stresses caused here by wind or other sources of intermittent loading. A kink can be removed in only one way: by cutting it out and then splicing the two wire ends back together. Solid copper wire that has less damage caused by bends or loops can be salvaged by stretching it to the yield point, which straightens it out. The same is true of high strength wire, but that same property makes it more difficult to stretch by simply pulling one end. Leverage of some sort may be necessary.

Installing the wire requires that it be pulled taut enough to avoid excessive sag at the midpoint, a condition typical of any horizontal wire. Even a vertical antenna wire must be sufficiently pre-stressed to prevent excessive movement in the wind. The loading force that results can be considerable, and it must be continually withstood during the lifetime of the antenna. In addition, there are the temporary forces caused by wind and ice. A very long antenna may require a spring-loaded termination at one end to compensate for temperature expansion and contraction of the wire. Posts, frames, masts, and their attachments may be quite simple for a short antenna, but can become elaborate for arrays and long wire installations. These can be design projects in themselves and are not treated here.

Insulators are required at each end of an antenna, regardless of what type of supports are used. These are available in either glass or glazed porcelain and are impervious to moisture. Types that absorb moisture should not be used. Ribbed construction for some of the insulators increases the surface path between the two ends, so that there is less leakage current when dirt and moisture buildup occur. A simple rod with a hole in each end works about as well for an antenna wire insulator except under worst case conditions. Plastic insulators are also available, but their use is somewhat limited by the relatively low tensile strength of plastics.

Ground

A good ground is essential for high sensitivity reception. The best choice for a ground connection is a metal cold water pipe, and if there is one inside the building and close to the receiver, the need for a second feed-through to the outside is eliminated. In such a case, an outside ground must still be provided in a separate circuit with a lightning arrestor for safety considerations. This will be discussed in the next section. A typical ground system using one outside pipe is shown in Fig. 2-3. The second feed-through is necessary when no inside ground is available.

A hot water pipe makes a questionable ground. This is because the plumbing at the hot water tank may not provide an all metal path back to the cold water inlet. A gas pipe should not be used as a ground because of safety considerations, even with presumably adequate protection from lightning or electrical circuitry. New housing may come with some of the piping in PVC, in which case there may be no inside ground at all. A problem can also arise if the receiver is located in a room well away from metal piping. In these situations, some other grounding method must be used.

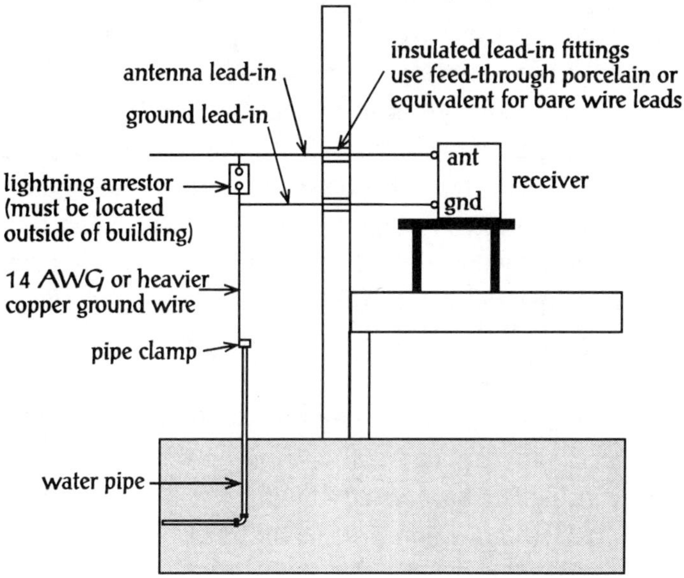

Figure 2-3: A ground system using an outside pipe and lightning arrestor.

A simple outside ground that works in moist soils is a one half inch or larger diameter metal pipe driven some three feet or more into the ground. Filling the pipe with water may help where the soil tends to dry out. Reception may improve if several pipes are driven into the ground at intervals of a few feet, or in a cluster, and are connected together. Another possibility is a metal plate buried in the ground. In dry climates, it may be necessary to supply several square feet of plate, at a depth at least equal to the principal plate dimension.

Lightning Protection

A lightning strike, as we all know, can injure or kill as well as do property damage. Protection of the antenna itself from lightning is impossible, but a lightning arrestor is a useful device for preventing the high current surge produced by a lightning strike from entering a building. The system is shown in Fig. 2-4. The lightning arrestor in the figure normally acts like an open switch, but when lightning strikes the antenna, arcing occurs, and it becomes a short circuit across the lead-ins to the building. The current surge from the antenna is diverted to the ground connection outside the building. This is the reason an outside ground is necessary even when the receiver is grounded separately inside.

The operation of the lightning arrestor depends upon two heavy metal electrodes spaced a few thousandths of an inch apart. The resulting air gap is the "open switch" referred to above. RF signal voltages and normal static are far less than the breakdown voltage across the air gap. The situation changes drastically when the antenna is struck by lightning. The voltage across the gap abruptly rises to a high value, reaching the breakdown voltage. The air in the air gap region ionizes, and a very low resistance path is established between the electrodes. This resistance is much less than the circuit resistance through the building and back, so the high current surge produced by the lightning is shunted directly to the external ground.

Suitable lightning arrestors are available at many hardware stores and electrical supply houses. It is also possible to construct one, using a spark plug re-set for about a one millimeter air gap. The narrow air gap should be protected by a plastic cup or some similar cover, fastened to the end of the plug. A pipe clamp can be used to connect to the opposite end. This is an arrestor in its simplest form, and a more elegant version can be constructed if desired.

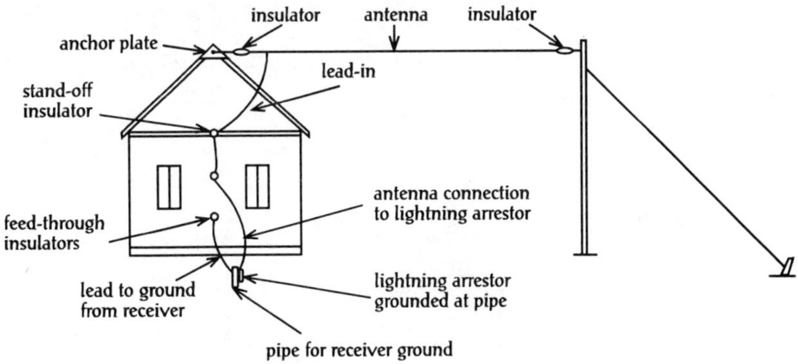

Figure 2-4: General arrangement for an inverted L antenna, with lead-in, ground for receiver, and lightning protection for building interior.

The preceding sections cover the complete set of components needed for an antenna-ground system. This brings the signal to the tuned circuits of the receiver, where the wanted signal is separated from unwanted signals and most of the noise.

Signal Selection

The previous sections have covered the antenna, ground and lightning protection aspects of crystal set receivers. The task of separating the signal of just one station from many and delivering the audio signal to the headphones are discussed below.

The voltage output from the antenna consists of the sum of the RF signals and noise that may be present. It is desirable to select a particular signal as effectively as possible and then apply that one RF voltage to the crystal detector. There are three ways in which to do this:

1) Pass through the wanted signal in terms of an increased voltage level.

2) Reject the unwanted signals in terms of decreased voltage levels.

3) Use a combination of both methods.

The first alternative is the method usually used. This is accomplished with an LC circuit—a combination of a coil and a capacitor—which is tuned to the frequency of the desired signal. The signal voltage at the resonant frequency is greatly increased, so that the wanted signal is delivered to the detector at a much higher level than noise and signals at other frequencies. However, noise at the resonant frequency cannot be separated from the signal, and it also reaches the detector.

The rejection method is used in a few special cases and also employs a tuned LC circuit. For this situation, the resonant frequency of one signal is rejected while signals and noise at other frequencies are passed. This type of circuit is commonly called a wave trap. It is used with at least one other tuned circuit to combine both methods of signal selection, producing one signal at the output.

More elaborate LC band pass filters and band rejection filters are seldom used in crystal sets because there is too much attenuation of the desired signal. This more than offsets their superior filter characteristics. Their application is in systems where amplification can easily compensate for signal loss. Resistance capacitance (RC) filters are never used at RF frequencies because of excessive signal losses. The practical methods of signal selection are discussed in detail below, as applied to each section of the crystal set, beginning at the antenna.

Antenna Tuning

Modern receiver design does not apply to the crystal set, as far as antenna tuning is concerned. Ganged tuning with one control was considered an essential feature for the home receivers built since the 1920's (superheterodynes). Such tuning was designed to optimize operation of the mixer and intermediate frequency circuits. Getting the last bit of radio energy out of the antenna was not the goal. To optimize the crystal set, however, the antenna and LC circuits must be tuned separately. There is an impressive improvement in crystal set performance when the antenna is tuned.

The most commonly used tuned antenna is the inverted L, discussed earlier. Both inverted L and vertical antennas are self-resonant at a wavelength which is roughly four times the length of the antenna. The relation is not exact because, among other things, the capacity to ground and the length of the lead-in also affect the resonance wavelength, as was shown earlier. The shortest antenna which is self-resonant in the broadcast band is about 150 feet in length, corresponding to 1650 Khz, and it is nonresonant at lower frequencies.

The antenna system resonant frequency decreases when the antenna coil of the receiver is connected between the antenna and ground. A variable inductance can be used to tune the antenna to lower frequencies, but it is usually more convenient to add a variable capacitor in series with the antenna coil and use it for tuning. A conventional 365 pF variable capacitor with a 250 μH coil can be used with a long antenna to tune through the entire broadcast band. A practical circuit for series-tuned antennas is shown in Fig. 2-5. Shorter antennas require additional inductance, capacitance, or both, to tune below about 800 to 1000 kHz, and an antenna of less than 40 feet is usually unsatisfactory for antenna series tuning. These very short antennas require that a tuning capacitor in the circuit be connected in parallel with the antenna coil instead of in series, so that the antenna coil itself is tuned. This is shown in Fig. 2-6. A practical arrangement for selecting either arrangement is given in Fig. 2-7.

In principle, the variable capacitors in Fig. 2-5 can be located at either the antenna or ground side of the coil, and this is true in practice for crystal sets which are not selective. However, a set that is very selective because of high circuit efficiency (high Q) requires that either the variable capacitor be on the ground side or that it be thoroughly isolated from the front panel controls. This is because stray capacitance to ground changes when a hand is brought near the panel, which shifts the antenna circuit resonant frequency. This hand capacity effect becomes very noticeable and annoying with sharply tuned circuits that are above ground potential. The easiest way to avoid the problem is to ground the capacitor rotor, as is done in Fig. 2-5.

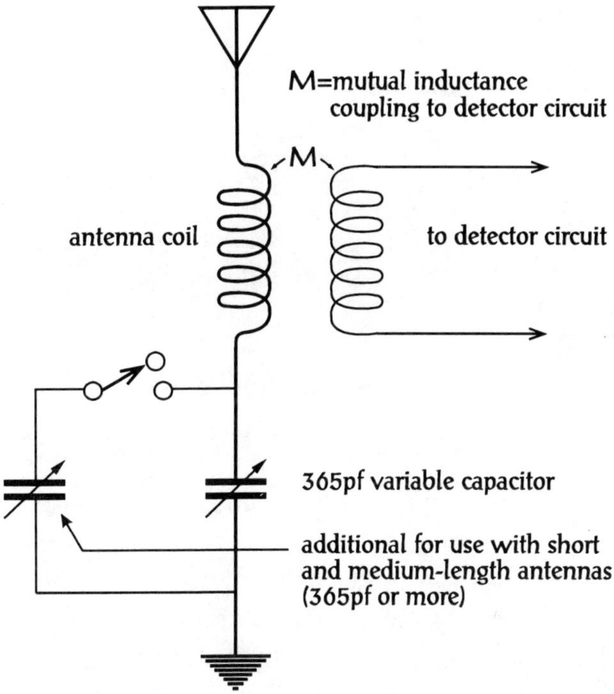

Figure: 2-5: Practical circuit for antenna tuning.

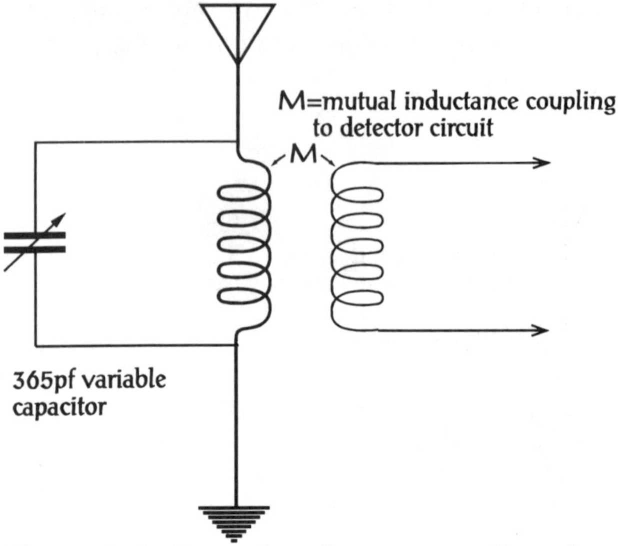

Figure 2-6: Example of antenna coil tuning with a short antenna.

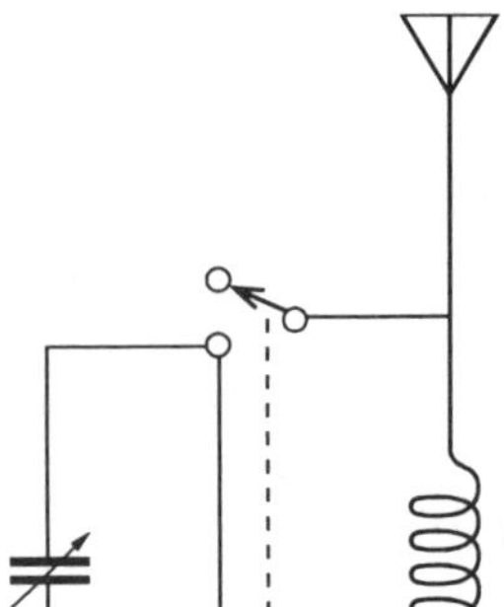

Figure 2-7: Switching circuit for selecting series or parallel tuning of antenna.

A good antenna tunes very sharply when coupling to the rest of the set is not too tight, and a narrow band, high amplitude signal is produced at resonance. This is why antenna tuning is so important. This is also why a straight wire antenna with a short lead-in is so desirable. Tuning also works well with more than one wire connected in parallel, provided that the wires are strung parallel and are of equal length, such as in an inverted L array. One or more wires strung irregularly around several corners may result in a loud signal, but they are untunable.

Detector Circuit Coupling and Tuning

Some of the simplest crystal sets place the detector directly across the antenna coil, as in Fig. 2-8. Either C_a or C_d is used for tuning. When both are present they interact, but there is sometimes an advantage for such a system because of its flexibility. In any case, the disadvantage of the Fig. 2-8 circuit stems from the fact that the detector and headphones series circuit across the coil has a relatively low impedance and loads the coil, decreasing the Q. Lowered Q means

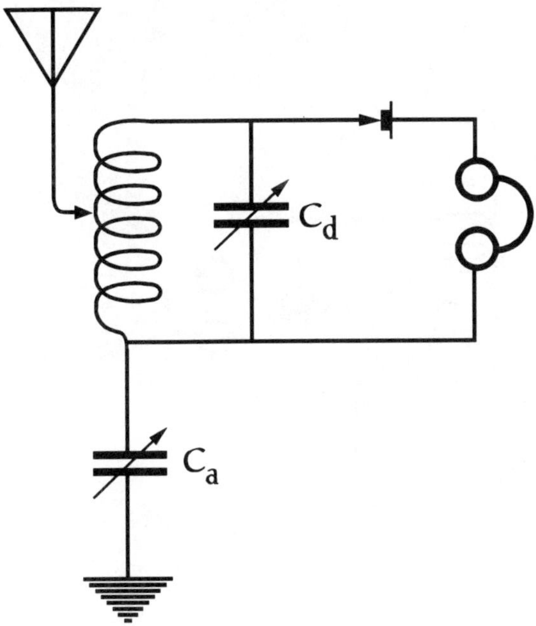

Figure 2-8: Simple form of crystal set with detector directly coupled to antenna coil.

lower circuit efficiency and broader tuning for both the antenna and detector. Overall selectivity is often poor, even when both capacitors are used and set for best operation.

Hand capacity with C_a on the antenna side, rather than as shown, is not a problem because of this lack of selectivity. One method of decreasing loading to improve selectivity is to tap the detector coil and load only a portion of it. The use of a tapped detector coil is discussed later.

A more complicated but more efficient system appears in Fig. 2-9. Here the detector is coupled by variable mutual inductance to the antenna coil, and this permits some choice between close coupling for improved signal output (high sensitivity) versus loose coupling for improved selectivity at

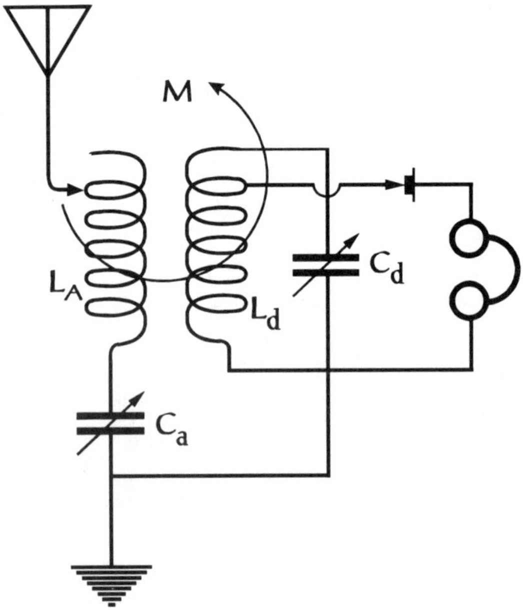

Figure: 2-9: Mutual inductance coupling between antenna coil and detector coil.

the expense of low sensitivity. As before, a tapped coil can be used to decrease detector loading. Standard values of inductance and capacitance can be used for L_d and C_d, and changing the tap location does not greatly affect the setting of C_d to maintain a given resonant frequency. Fig. 2-9 can be used as the basis for some excellent crystal set designs.

Special coils are a desirable alternative to commercially available products. In general, L_d and C_d may be chosen in terms of the resonant frequency formula,

$$f_r = \frac{1}{2\pi\sqrt{L_d C_d}}$$

where f_r is the minimum resonant frequency of the tuned band in Hertz, L_d is the inductance in Henrys, and C_d is the maximum value of the variable capacitor in Farads. In more practical form the basic relation is

$$f_r = \frac{159{,}155}{\sqrt{L_d C_d}}$$

where the units of measurement are

$[C_d] = \text{picofarad}$
$[f_r] = \text{kilohertz}$
$[L_d] = \text{microhenry}$

Another method of coupling the antenna coil to the detector is by means of small capacitors, as shown in Fig. 2-10, and is called capacitive coupling. This method was used to some extent in the early days of radio, but later lost popularity in favor of the inductive coupling methods described in preceding paragraphs. There seems to be no particular advantage in this method to offset the requirement for additional parts. Indeed, at least one early day textbook stated that capacitive coupling was less selective than the inductive method. One or both coupling capacitors can be variable, and the return side capacitor has been omitted in some versions of the circuit without notable differences in performance. In that case, the return circuit depends upon stray capacitance. It is necessary to reduce mutual inductance between the two coils as much as possible here, or operating peculiarities may result.

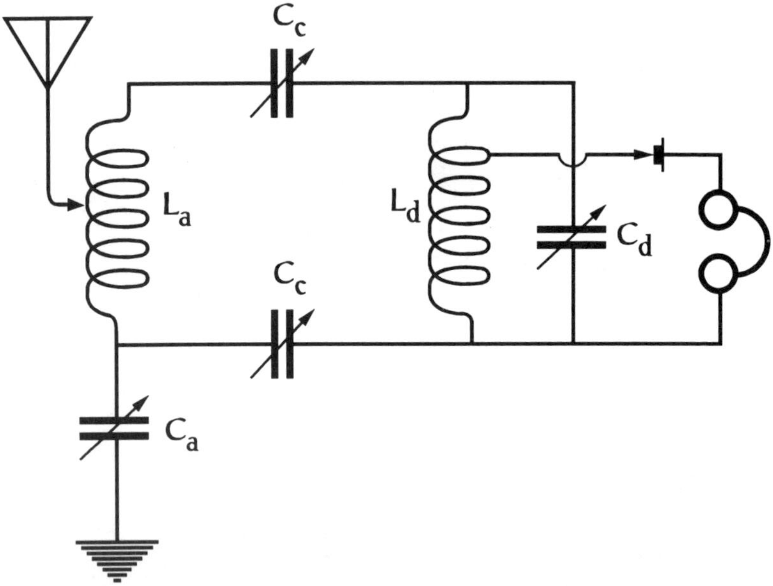

Figure 2-10: Capacitance coupling between antenna coil and detector coil.

Tuned Transfer Circuits

The possibility exists of improving the performance of such circuits as the one in Fig. 2-9. The approach is to change the way energy is transferred from the antenna to the detector to an arrangement where a transferring circuit is tuned. In theory, only the desired signal energy is carried over from the antenna coil to the detector. At least two types of circuits have been employed, both dating back many years.

The most straightforward and easy to understand method is tuned link coupling, one version of which is shown in Fig. 2-11. Here L_a and L_d are isolated from each other as much as possible. Energy is transferred from one to the other by means of $L_{\ell 1}$, $L_{\ell 2}$, and C_ℓ. In another form of this circuit, $L_{\ell 1}$, $L_{\ell 2}$, and C_ℓ are all connected in parallel. Variations are possible, such as grounding one or both floating circuits. Detailed performance information on link coupling for crystal sets is lacking, but it has been much used in the past. One of the most highly regarded receivers of the World War I period was the American Marconi Type 107A, and it used tuned link coupling with a crystal detector in one of the system arrangements.

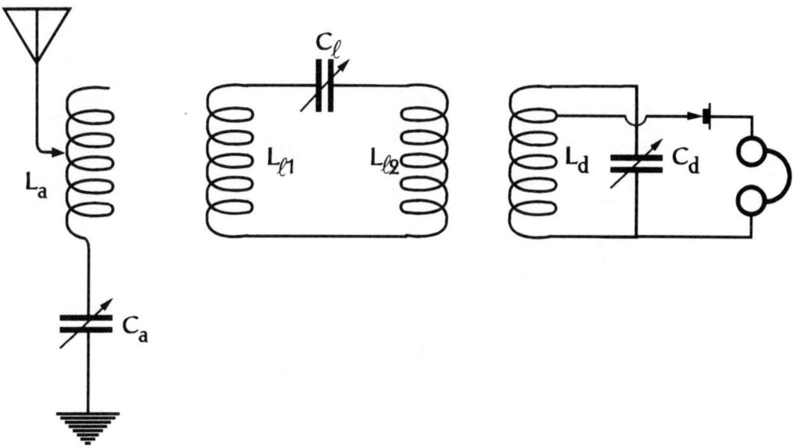

Figure 2-11: Link coupling between antenna and detector circuits.

The second type of circuit appears in Fig. 2-12. Good results have been reported for some variations on this basic arrangement, where energy from the antenna coil is transferred by a single tuned coil located between the antenna and detector coils (all in-line). The Q of the system as a whole is very high when good coils are used, and selectivity can be surprisingly high without undue loss of sensitivity. One of the drawbacks of this approach is related to that high Q circuitry. Such systems are loosely coupled by necessity, and so become extraordinarily sensitive to mutual inductive and stray capacitive effects between the components and to the surroundings. The operating peculiarities which sometimes result can only be described as bizarre.

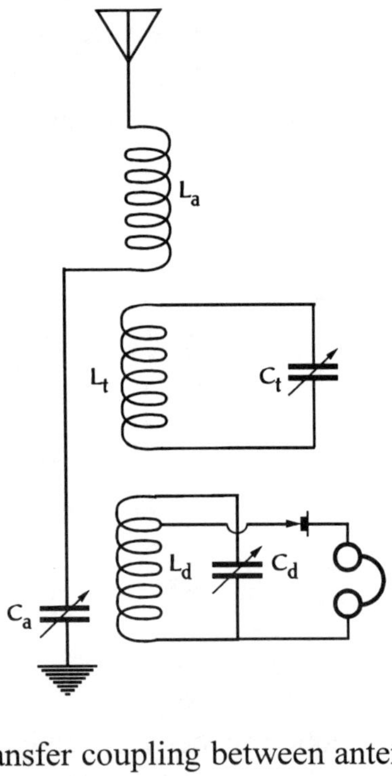

Figure 2-12: Transfer coupling between antenna and detector circuits using a third tuned circuit.

Wave Traps

Wave trap circuits were quite popular at one time, when it was difficult to design sharply tuned receivers. They still have their applications with crystal sets, where selectivity is often a problem. A wave trap is a simple form of filter and can be used in the same way as band pass or band rejection filters are used. An acceptor wave trap passes its resonant frequency while a rejector wave trap rejects that frequency.

A form of acceptor wave trap which improves the performance of a simple crystal set, such as the one in Fig. 2-8, is shown in Fig. 2-13. Here the trap L_tC_t is coupled by mutual inductance to the antenna coil to boost the output at the resonant frequency. This has sometimes been called absorption tuning. Depending upon the amount of coupling M, there may be noticeable interaction between C_t and the other capacitors when tuning the set. In principle, this approach can be used with any tunable circuit.

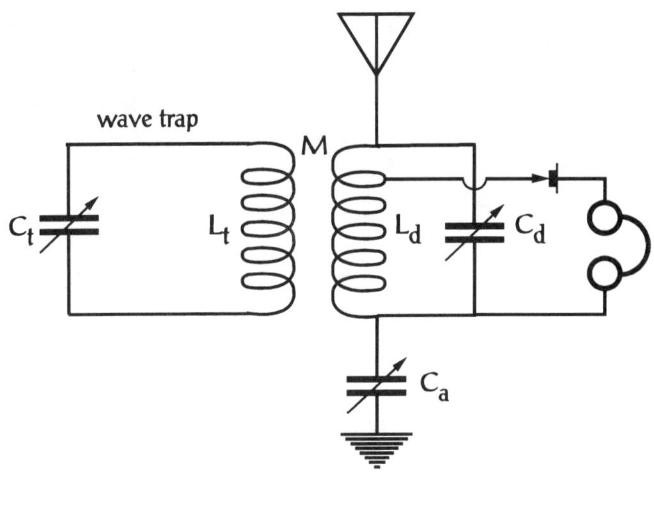

Figure 2-13: Acceptor type wave trap for increasing signal output to detector.

The rejector wave trap is most commonly used in the antenna-ground circuit to tune out a strong unwanted signal which is interfering with reception. This arrangement is shown in Fig. 2-14A. The version in 2-14B is often preferable because the amount of coupling can be adjusted for a particular set of operating conditions. Some other wave trap locations are shown in Fig. 2-15.

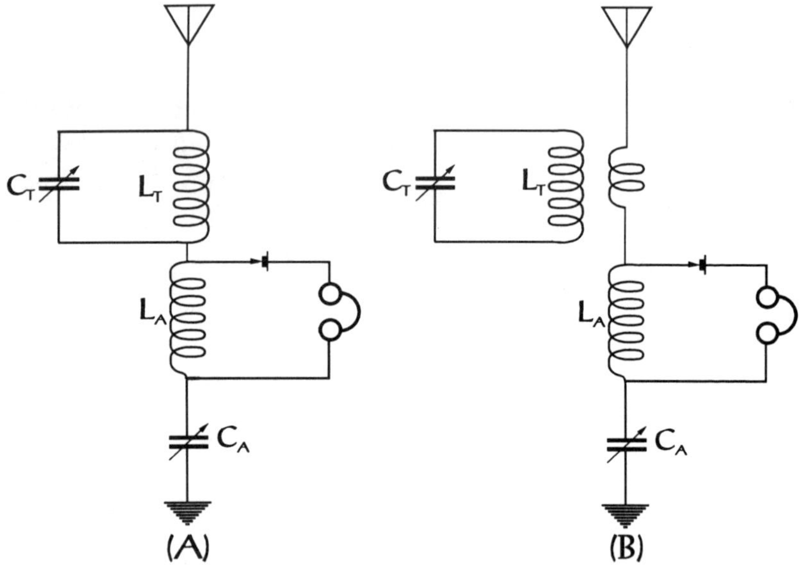

Figure 2-14: Rejector wave trap designs for antenna circuit applications.

Wave traps all have a common disadvantage; they interact with the associated tuned circuits. This is not a major problem with acceptor type circuits such as Fig. 2-13, but rejector circuits seriously affect the tuned circuit characteristics. The concept of tuning to a weak distant station and then tuning a rejector wave trap to eliminate a second interfering station is an impossible dream. The resulting change in the system resonant frequency is so great that the desired station is no longer in tune, and if it is too close to the wave trap resonant frequency it may be lost altogether.

The only practical usage is to preset the trap at the frequency of an interfering station and leave it there while operating in that general portion of the band. Also, a rejector trap is useful only for strong local signals that have a broad bandwidth, because the trap itself is not selective enough to discriminate between two signals of comparable amplitude and close together in frequency. In many situations it is better to optimize set selectivity than to depend upon wave traps, but special cases do arise where a rejector trap may be helpful.

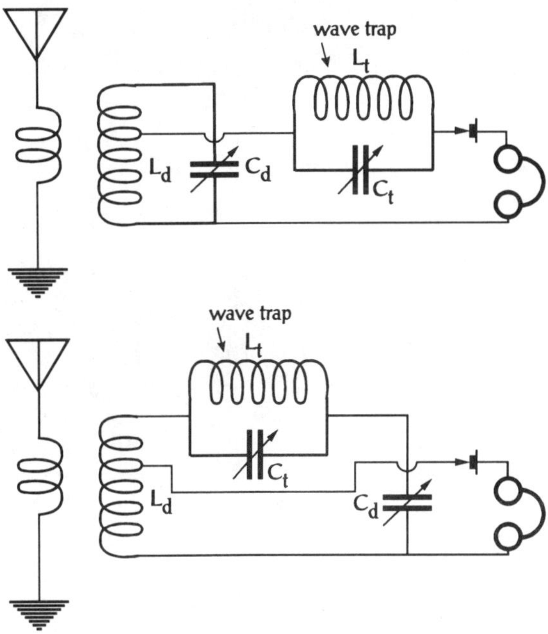

Figure 2-15: Rejector wave traps in detector circuits.

Signal Detection

The signal that reaches the detector is still a radio frequency voltage, regardless of the amount of signal processing between the antenna and the detector input. It consists of an RF carrier modulated by an audio waveform. Fig. 2-16 shows a typical signal at the detector input. Here, the envelope is the modulating audio signal, and the positive and negative components of the envelope are 180° out of phase. Any device sensitive only to audio frequencies would display zero output if connected at this point, since the two envelope components cancel at any given time. A peak-to-peak RF voltmeter would show either a constant or fluctuating voltage, depending on the signal and meter inertia, unless the carrier were unmodulated. In that case, a constant RF voltage output would result.

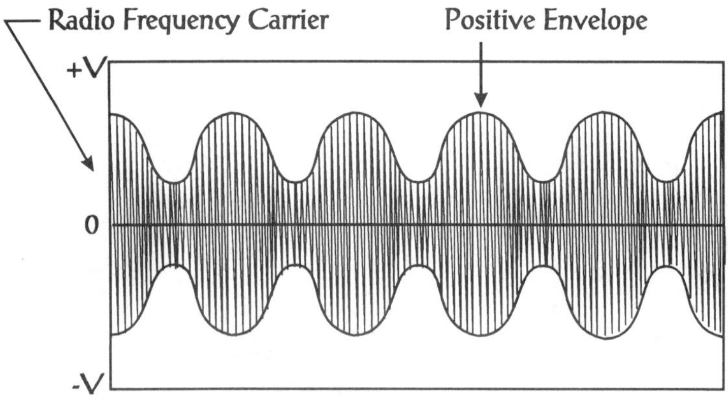

Figure 2-16: Radio frequency signal at input of diode detector.

The Detection Process

Crystal detectors fall into the diode detector category, and a characteristic common to all diodes is the property of rectification. By definition, a rectifier is a device which passes electrical current more easily in one direction than the other. An ideal rectifier has zero resistance to current flow in its forward direction, and infinite resistance in the opposite (reverse) direction. Therefore, with an ideal rectifier, the input voltage shown in Fig. 2-16 produces zero current through an output-side load whenever the input voltage is negative. The output voltage across the load must appear as in Fig. 2-17 since there is current flow whenever the input voltage is positive, with a corresponding voltage drop across the load.

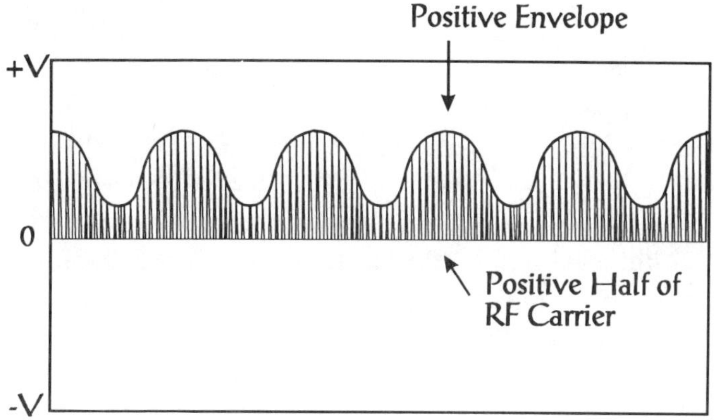

Figure 2-17: Idealized diode detector output after half-wave rectification.

A test circuit for measuring diode characteristics is given in Fig. 2-18. It is convenient to use a commercial model signal generator with a RF output which can be either modulated by an audio sine wave or unmodulated. Meter V_{in} may be available with the generator to display the RF output voltage. The characteristics of the crystal are determined when the switch is closed to shunt out the load impedance. In that case, the output current produced by unmodulated input voltage V_{in} is measured by AC millimeter i. Behavior under various output loads is determined by opening the switch and reading the output voltage. Input voltage is kept in the order of one volt or less for the type of measurements which are applicable here.

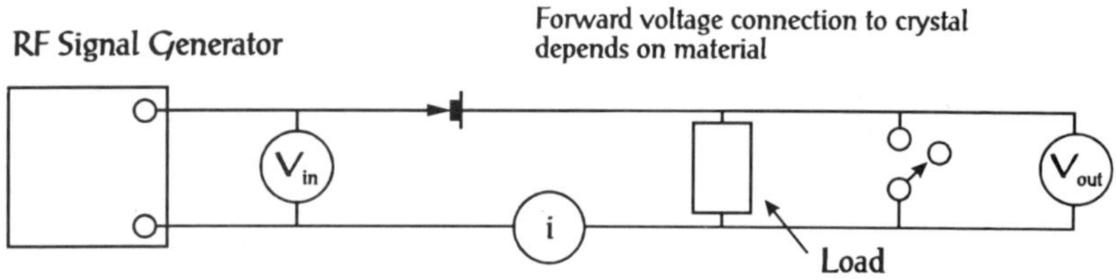

Figure 2-18: Test circuit for measurement of diode characteristics.

The output current waveforms for ideal and real solid state diodes are shown in Fig. 2-19, with zero impedance load. The ideal detector permits no negative current flow. However, even the best real diodes allow some flow in the reverse direction, because they do not fully rectify the current. That is why there is a small negative current flow in Fig. 2-19B. This current subtracts from the forward current, on an averaged basis, so that the net audio output current i is lower than for an ideal diode. Nevertheless, a good diode will show a substantial net forward current output.

Direct current measurements also provide a great deal of information about diode detectors. In fact, early day investigations of crystal detector properties depended upon this type of measurement. The circuit, shown in Fig. 2-20, is quite similar to Fig. 2-18, but a variable DC power supply is used, with some convenient provision for reversing the output polarity. Measurements are made with zero impedance loading. Commercial diodes are evaluated over a wide range of positive (forward) and negative (reverse) voltages to fully determine their characteristics, but here the necessary DC input voltage ranges over only a volt or so in magnitude, with corresponding current output in the order of milliamperes or less. With this arrangement, the voltage-current curves for ideal and real diodes are given in Fig. 2-21.

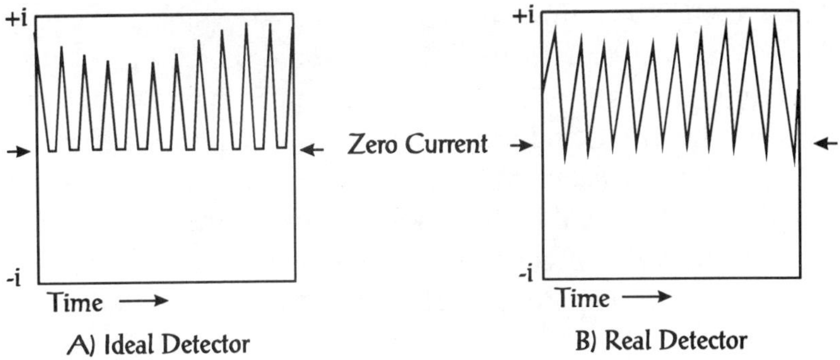

Figure 2-19: Output current characteristics of ideal and real diode detectors with zero impedance loads.

The ideal detector, in Fig. 2-21A, conducts perfectly when the input voltage is positive, because it has zero internal resistance in the forward direction. No current is passed when the input voltage is in the reverse direction because the internal resistance becomes infinite. Should an ideal diode be measured with a load resistance at its output, the output current becomes $i_{out} = V_{in}$ /(load resistance) in the forward direction and zero in the reverse direction.

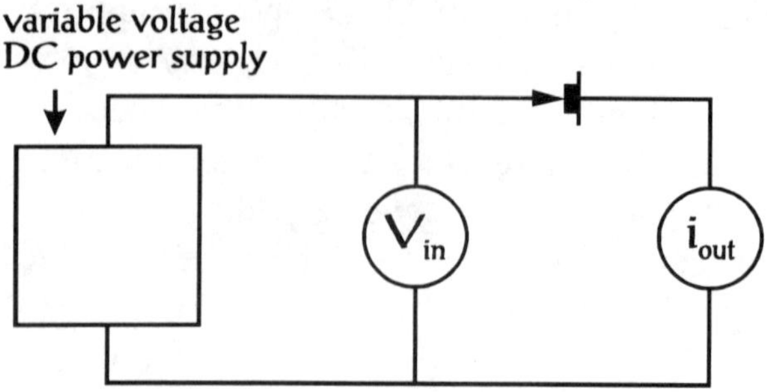

Figure 2-20: Test circuit for measurement of diode DC characteristics.

A real diode has a finite internal resistance, which prevents it from conducting perfectly in the forward direction. Consequently, i_{out} rises as shown when V_{in} is increased in the forward direction, and is given by $i_{out} = V_{in}$/(internal resistance + load resistance). The internal resistance does not remain constant as the voltage increases, so the forward current versus voltage curve is not a straight line, although it usually becomes approximately straight at angle θ in the

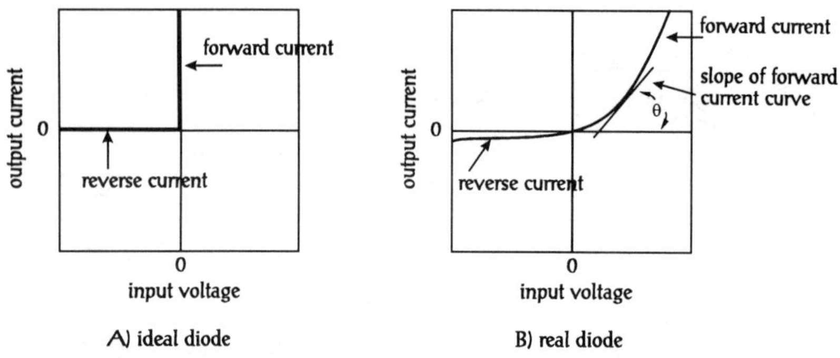

Figure 2-21: Current-voltage curves for ideal and real solid state diodes.

figure as the voltage is increased. In the reverse direction, the diode resistance is much higher but still finite, so a low, gradually increasing reverse current results as the negative voltage is increased in magnitude.

The curves in Fig. 2-21 are applicable at RF frequencies as well as for DC. The net current which determines the modulating signal output is the difference between the forward and reverse currents for a given peak to peak RF voltage. The forward current results from the positive peak voltage. The reverse current results from the negative peak voltage. This was shown for RF voltages in Fig. 2-19B, where the presence of the small negative RF voltage actually decreases the modulating signal output. The difference between the two peak current magnitudes in 2-21B, not their sum, is the measure of how large the audio output may be.

The difference between the voltage envelopes in Fig. 2-19B is the maximum available audio output from the detector. This net voltage is readily convertible to audible sound by means of earphones, without significant additional signal processing, as is discussed in the next section. First, however, it is necessary to consider the current-voltage curve in further detail, and to describe a property that can be used in some cases to increase the available signal output.

The internal resistance of a crystal decreases as the forward voltage across it increases. This was mentioned above, and it is another demonstration that active substances do not follow Ohm's Law. As a result, more current is produced when the resistance is lower; that is, sensitivity is higher when the slope θ is greater. On the other hand, if a point near zero voltage were chosen, where the slope is less, sensitivity is greatly reduced. By imposing a DC voltage on the crystal, even though the RF signal may be small, we can operate the diode where θ is larger (current is higher). Hence, sensitivity to a weak RF signal is increased. One way to do this is shown in Fig. 2-22 and is called biasing the crystal.

Current-voltage curves differ depending upon the material of the diode. The best operating point θ in many cases is so close to zero that no bias is necessary. In other instances, over a volt of bias is needed for best sensitivity. Carborundum is a classic example where several volts of bias may be necessary.

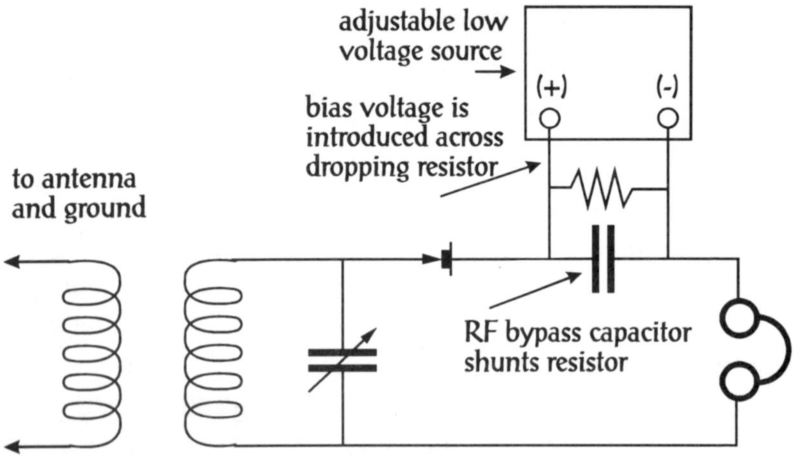

Figure 2-22: Method of biasing a crystal detector by introducing a constant voltage into the detector circuit voltage loop.

The Audio Output

The electrical and mechanical characteristics of high impedance magnetic diaphragm earphones are usually sufficient to provide a satisfactory output load for a crystal detector. A typical output schematic and its equivalent circuit are shown in Fig. 2-23. The inductance of the coils, wound on magnetic cores, act as chokes to stop flow of RF current. The self capacitance of the coils is usually sufficient to store RF energy during each half cycle of the carrier voltage. Sometimes a capacitor C may be added as shown in the figure to improve upon this situation. In any event, the net result is a filtering action which permits passage of an average signal level that closely approximates the RF envelope. See Fig. 2-24. The diaphragms of the earphones respond readily to this signal to produce the required sound. At the same time, they are mechanically insensitive to any RF ripple still present at the output, because they cannot vibrate at such high frequencies. As a result of both electrical filtering and mechanical inertia, the output of the detector is converted to sound.

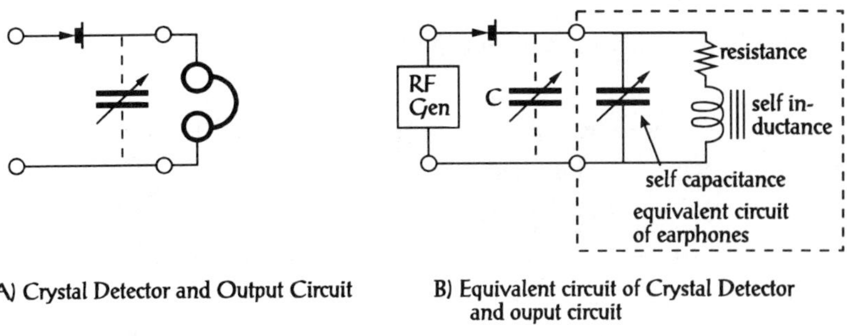

Figure 2-23: Crystal set output circuit.

The high impedance magnetic diaphragm earphones described earlier are the most commonly used type for crystal sets. Other types are available and some have been used. These may or may not perform well and in some cases require additional circuitry to provide a good impedance match to the output of the receiver. The most often used of these are piezoelectric earphones, often called crystal earphones. These depend upon the expansion and contraction of a material which changes its dimensions when an electric field is applied across it. This change in dimensions drives a diaphragm to produce sound.

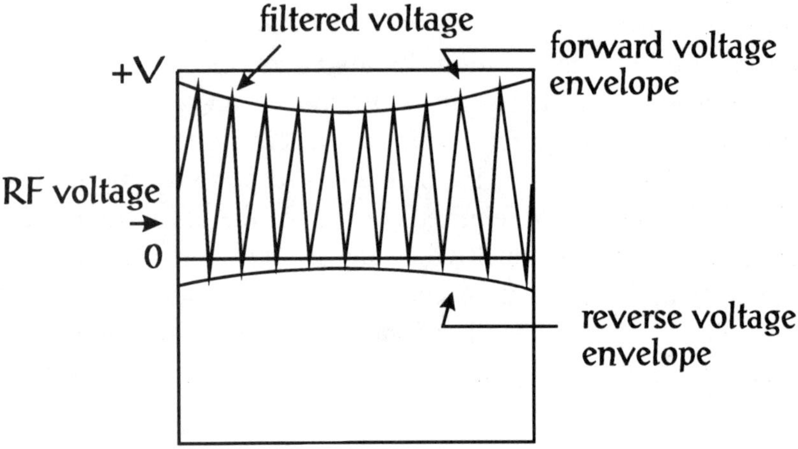

Figure 2-24: Output voltage at earphones.

The impedance across the input terminals is significantly higher than even the best quality magnetic diaphragm units, so there is less loading of the detector circuit. However, results are satisfactory only with better quality units such as the Clevite "Brush" type with impedances as high as 200,000 Ohms at 1,000 Hz. In any case, a shunting resistor is necessary across the earphones to prevent DC voltage build-up, because the impedance of the earphones is almost entirely capacitive. Cheaper crystal phones require a lower shunting resistance to decrease distortion and improve bass response. This increases detector circuit loading and tends to eliminate the advantage of using them.

Loudspeaker operation of a crystal set is not usually practical, since the available power is so small. There are a few exceptional cases, often deriving from the fact that there is a high power transmitter nearby, and that the receiver has been designed for high sensitivity. Results have been reported when a small permanent magnet speaker is coupled by an impedance matching transformer to the detector output. Better performance is likely by converting an earphone to a speaker by adding a conical horn at the ear piece; one of the early day horn speakers using that same approach may be employed. Usually, such a system is not very practical and is treated as an oddity or stunt.

BIBLIOGRAPHY

Allied Radio Corporation Catalog 140, 1955.

American Radio Relay League, The ARRL Handbook for Radio Amateurs. 70th Edition, 1993.

Anderson, Philip N., The Crystal Set Handbook. The Xtal Set Society, Inc., Lawrence, Kansas, 1994.

Bucher, Elmer E., Practical Wireless Telegraphy. Wireless Press, Inc., New York, 1917.

Chemical Rubber Publishing Co., Handbook of Chemistry and Physics 35th Edition. (Also both earlier and later editions).

Edwards, K. E., Radios that Work for Free. Hope and Allen Publishing Co., Grants Pass, Oregon, 1977.

Ghirardi, Alfred A., Radio Physics Course. 2nd Edition, Murray Hill Books, Inc., New York, 1932.

Green, Charles, All About Crystal Sets. Allabout Books, Fremont, CA, 1984.

Grover, Frederick W., Inductance Calculations. D. Van Nostrand, 1946. Reprinted by Instrument Society of America, Research Triangle Park, North Carolina, 1973.

Pierce, George W., Principles of Wireless Telegraphy. McGraw Hill Book Co., Inc., New York, 1910.

Robertson - Cataract Electric Co., Radio Equipment and Supplies. Catalogue No. 22, 1922-1923. Reprinted by The Vestal Press, Ltd., Vestal, New York.

Rosen, Marvin and Phil Anderson, communication in The Xtal Set Society Newsletter, Vol. 4, No. 1, Jan 1, 1994, p.5. (concerning suggested use of a transistor without external power).

Terman, Frederick Emmons, Radio Engineering. 2nd Edition, McGraw Hill Book Co., Inc., New York, 1937.

Thackeray, Desmand, The Carborundum Story. The Old Timer's Bulletin. Vol. 27, No. 3, Nov. 1BL6, p.10.

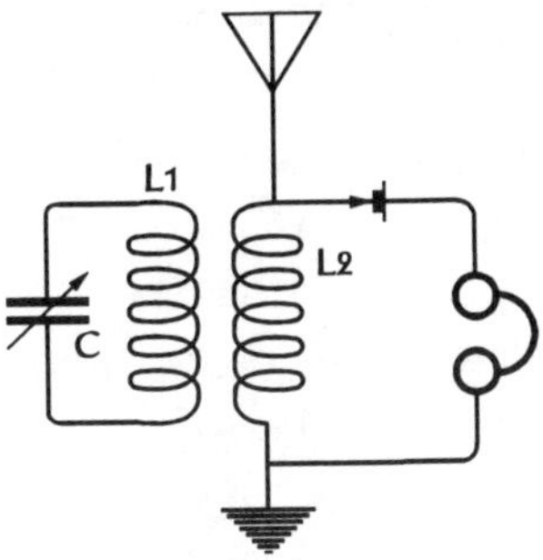

CHAPTER 3

CRYSTAL SET DESIGN

Introduction

There was a time during the early days of radio when crystal set design was part of the main-stream technology that we now call electronic engineering. The use of crystal detectors and several other detection methods evolved at about the same time, using similar circuits and components. In fact, some of the receivers of that period had provisions for two or more detector types on an interchangeable basis. Significant improvement of the first crude vacuum tubes changed all of this, making the crystal set obsolete, first as a primary method of signal reception, and, after many years, even for standby or emergency use. After the early 1920's, design methods diverged from crystal set era technology as electronics developed along other paths.

Present day electronic engineering texts cover crystal set design only where by coincidence there is still some common ground. In fact, there are now problems in crystal set construction which did not exist during the early decades of this century. This is because some modern components do not work well in crystal receivers, and early day versions are no longer produced. Situations arise where it is necessary to either modify a modern part, or build one on a do-it-yourself basis, all to provide something that could have been bought off the shelf before 1930 or 1940.

This chapter gives the information needed to design radios using crystal detectors, and there are several approaches to this process. Among other possibilities, the designer may want to surpass present day performance with something new, reconstruct a circuit given only partial information, or analyze an early day receiver to construct a modern replica. These and other alternatives are covered in the section on design criteria. It will be shown that some properties and functions are inevitably emphasized at the expense of others, and that there are decisions to be made in that respect before proceeding further.

A following section describes circuit options that are available, after the design goals have been established. There is no one "best choice" or universal solution. Once a basic type of receiver is chosen, there are numerous possibilities for changing or adding features. These are covered in the final section on circuits and systems with a proven track record for good performance. Also worth consideration, and included, are some complete receiver systems which are so unique in one or more respects that they stand on their own. These are presented for either direct application, or for comparison with other options.

Design Criteria

Design depends upon application. This is just as true for crystal sets as it is for anything else built by human hands. Therefore, it is important to choose the desired application requirements before proceeding with a particular design. These usually include some of the following design parameters:

- Sensitivity — a "must" for long distance reception
- Selectivity — for separation of interfering signals
- Size — is compactness important?

- Simplicity — straightforward construction, convenient operation
- Appearance — "traditional" or modern
- Antenna — compatibility with what is available

The choice of criteria is not as easy as it first appears. A desire to meet two or more requirements on the above list is likely to produce conflicting conditions which can be met only by compromise. The classic example of this problem is the trade-off choice between sensitivity and selectivity. The two properties are so completely interdependent that they will be discussed together. Other criteria will be covered separately, with some comment on interdependence where necessary.

Sensitivity and Selectivity

Is your primary goal distant reception or local coverage without interference? Superficially, this is a clear-cut choice between a receiver optimized for best possible sensitivity or for maximum selectivity. However, as a typical example, many people live within twenty or thirty miles of several broadcast stations, and just one of these can produce a wide interference band on a low selectivity set. Any advantage of high sensitivity is lost within this band. In fact, a high sensitivity, low selectivity set is a useless artifact in metropolitan areas with many transmitters, as far as distant reception is concerned. On the other hand, emphasizing selectivity decreases sensitivity, making long distance reception less likely, so it is necessary to consider the interdependence of these properties.

There are mathematical formulas for selectivity and for sensitivity, expressed as a ratio of output to input signals.

These depend upon each type of circuit. The formulas are elaborate except for very simple circuits, with significant approximations required in some cases. For practical purposes it is sufficient to state that a relationship between selectivity and sensitivity almost always exists where increasing selectivity results in decreased sensitivity and vice versa. The simplest method of investigating that relationship is by direct experiment.

Adjustment of component values and/or coupling for a given receiver establishes a certain amount of trade-off between selectivity and sensitivity. However, the relationship changes when the circuit itself is changed, leading to a new interdependence and a new choice of trade-offs. Adjusting parts values or circuit coupling for a given receiver may not be enough to obtain good results. It may be necessary to change the entire receiver system. Some specific examples are given below.

It is possible to design a selective tuner which will separate local stations quite effectively, but the resulting signal levels are low. This type of system has been used with audio amplification to improve the output level, but the result is not a crystal set. In certain circumstances, a selective tuner driving earphones without amplification has been advocated as an excellent high fidelity AM receiver. Only local reception is practical at best.

A multicomponent narrow band pass filter and amplification may be used for these high fidelity applications, rather than the tuning arrangements described in the previous chapter. Although antenna tuning would improve reception, it is unnecessary and omitted when amplification is used. Details concerning band pass filter design are given in most electronics textbooks and are not covered here. Signal attenuation of the pass band signal through the filter is too great for crystal set application.

The preferred approach for good crystal set selectivity is to use a combination of antenna tuning with one or more of the tuned circuits discussed in Chapter 2. The number of possible combinations of these is very large, especially when including the variations of each circuit which can be considered.

As a specific design example, consider what must follow after selecting a system with moderate selectivity and sensitivity for use outside of station-congested areas. A basic arrangement is shown in Fig. 3-1. This is a good system from which to start, but by no means the only one. Some comments can be made about this and similar circuits, based upon accumulated empirical results. First, at least two tuned circuits are necessary to reach a reasonable degree of selectivity, unless a very insensitive receiver is acceptable. Second, for application here, the "bigger is better" approach to improved coil performance reaches the point of diminishing returns for single layer air core solenoid diameters greater than about 3 inches, and for wire sizes lower than 20 or 22 AWG, i.e., large diameter wires. Also, variocoupling should be used as shown in the figure, at least on a preliminary basis, to determine the amount of antenna detector coupling for the specific application. It may or may not be desirable to provide for variable coupling as one of the controls on the finished receiver. All of these limitations should be considered as guidelines only, which may be exceeded under some circumstances.

The component values shown in Fig. 3-1 for the antenna coil and antenna tuning capacitor are representative for a medium to long length inverted L antenna and broadcast band reception. One or both values may be changed for short or for very long antennas.

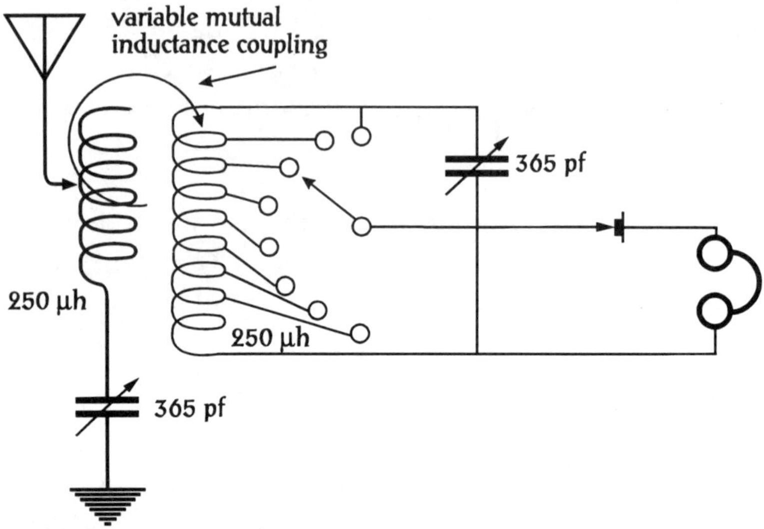

Figure 3-1: One basic type of selective circuit.

Tapping the detector coil is important. It takes advantage of a very unusual situation: both sensitivity and selectivity improve over a limited range when the taps are used. First, the signal output rises as the fraction of the detector coil length that is loaded by the detector decreases, due to better impedance matching between coil and detector circuit. At the same time, selectivity improves as detector loading of the tuned LC circuit decreases. At some point, sensitivity reaches a maximum and then decreases as the switch is advanced further, while selectivity continues to improve. Just where the maximum occurs depends upon the rest of the circuitry, in terms of component geometry and parts layout, as well as the exact circuit design, because the system is influenced by stray capacitances. It is not a very critical adjustment, but it would be well to use at least five or six taps, if a continuous sliding contact is not chosen as an alternative.

Tapped windings or sliding contacts were a normal feature of early day receivers, but they fell out of favor in more modern designs. Many crystal set circuits simply connect the detector to the upper end of the coil and good performance results, but the improvement using a tap or slider at the detector coil should more than compensate for the added system complexity. It is used here to bring the selectivity up to a moderate level without sacrifice of sensitivity.

Tapped windings or sliding contacts were also used in other tuned circuits in early day receivers, and there was often some benefit gained by varying the ratio of inductance to capacitance at a given frequency in a tuned circuit. This made it possible to adjust for the most satisfactory operation. The tuning of such a receiver demands real skill for effective use, especially if there are two or three such circuits, because the same signal can be tuned-in over a wide range of dial settings. These options have not been used here because the added benefit would be unlikely to make the inconvenience worthwhile. Such an addition would be worth considering for a more extreme design, to trade convenience in favor of a better combination of selectivity-sensitivity.

The type of coil which is used is not too critical for the present example, provided that it is adaptable for tapping or for a sliding contact, and that it is a high Q design. Air core single layer solenoids are the usual choice. Basket weave coils are another good example, and a short single layer solenoid wound on a ferrite core works well.

Some care should be used in selecting other components when considering selectivity and sensitivity. Most 350 or 365 picofarad air dielectric variable capacitors, modern or early day, are satisfactory for the broadcast band. The 140 pF size is standard for short waves. However, modern subminiature units that depend in part on a solid dielectric may not qualify. The Q ratings of some of these are less than desirable for an optimum choice.

Size

This is a parameter which may be all-important to some and of no consequence to others. As an example, during the early to mid 1920's, radio oriented publications gave considerable attention to the "smallest" crystal set, and to tiny receivers housed in unusual containers. A typical case was a crystal set built into an earphone shell, or built upon the outside of a working earphone, which was used at its output. Performance must have been poor, and the fad is hard to understand in view of the necessity for external antenna and ground, which eliminated any possibility of being portable while operating. It is evident that size is the dominant design choice for replicating a set of this type.

A more contemporary application is the design of improved miniature receivers as replicas of the little sets advertised for many years as toys or novelties. Miniature versions of coils and variable capacitors are available now that are adaptable for such an application.

Size is also linked to some degree with sensitivity and selectivity, because size affects coil Q, and to a lesser extent the Q of variable capacitors. Small coils are inefficient. Further, if a hand capacity problem arises, isolation of the offending components may require a larger layout. Thus, if both size and sensitivity-selectivity are important, a design trade-off is necessary.

Simplicity

Straightforward circuitry, layout, and packaging become important when designing for production. This is undoubtedly a factor in the case of the small businesses that advertise crystal set kits and fully assembled receivers for sale. This type of enterprise cannot afford to produce a receiver requiring excessive parts and extra labor for their assembly.

Simplicity is also a factor that the beginner should consider. It is wise to start with a simple project at the beginning of the learning process, in terms of knowledge of electronics, construction technique, and crystal set limitations.

If you wish to build a tried and true receiver design, rather than experiment, then avoid elaborate high Q systems. High Q, as applied to over all system efficiency, is quite possible with some of the sets using several tuned circuits. This is in spite of the fact that power is consumed in circuit losses and at the earphones. However, high Q usually means increased circuit complexity. Also, stray capacitances and unwanted mutual inductances in these systems may produce unexpected and sometimes peculiar characteristics. On the other hand, the dedicated experimenter may enjoy leaving simplicity out of his circuit limitations for this very reason.

Appearance

Is the new project a reproduction of an early day radio? If so, the restrictions on appearance are rigid and unyielding. Components, materials, and layout must conform exactly with the original set. There is no receiver design phase at all, though tooling up for construction may itself prove to be a major challenge.

The construction of a replica may present a different situation. The term replica has been applied to a wide range of possibilities; near reproduction quality is at one extreme while traditional external styling of a modern internal system is at the other. In the special case of crystal set construction, the tendency is to retain a traditional, i.e. early day, flavor for the appearance of the front panel and major components. In contrast, you may decide to try a new design and use modern components throughout, with adaptations where necessary for the special requirements of crystal sets. The most significant adaptations are usually associated with the coils.

There is no unwritten law requiring the separation of modern and early day styling, but some critics may object to such cases as the use of a three inch diameter bakelite dial for tuning the antenna, together with a broadcast band calibrated slide rule dial assembly for the detector circuit. The old adage about beauty being in the eye of the beholder can be applied here, unless conformity with the past is an important factor in the design.

Antenna Compatibility

The antenna circuit depends upon the type of antenna that is available when optimum performance is wanted. The importance of antenna tuning has been repeatedly emphasized for that condition when a tunable antenna is used. A medium to long length antenna requires a variable capacitor in series with the antenna and antenna coil. A very short antenna requires that the capacitor be connected in parallel with the coil. In some transitional cases it may be necessary to provide switching to accommodate both possibilities. See the circuit shown in Fig. 3-2.

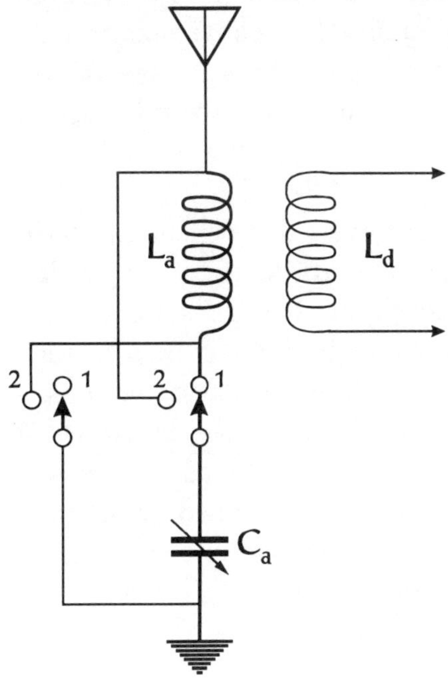

Figure 3-2: Antenna tuner switching ciruit for moderately long antennas.

A tunable antenna may be impractical in some cases, such as on a small lot or in an apartment building. A wire strung around a room is not tunable. In these cases, a different approach to receiver design is needed. Consider adding only one more tuned circuit ahead of the detector. Such a set usually tunes rather broadly, but sensitivity is good. This is a necessity for the low level signals that the antenna produces.

Design Options for Applications

The design criteria discussed earilier determine the circuitry which can be used for a specified application. General guidelines are helpful in making a selection, and these are developed in following paragraphs. It is convenient to categorize crystal sets in order to do this. There are several reasonable approaches to categorization, and a two-fold method is used here.

First, there is a large class of receivers that depend primarily upon variable capacitor tuned circuits for frequency selection. These are easily discussed, and it is possible to make some general statements about their relative performance. This class is presented in order of increasing complexity—the number of tuned circuits present—which is an important consideration when a simplicity compromise is needed. The variable capacitors in these tuned circuits can be replaced by fixed capacitors, with variable inductances (variometers) used to do the tuning. Alternatively, a mixture of variable capacitors and inductances can be used. Generally, these systems can be classified with those above.
The rejector wave trap is treated as a special situation.

Specialized variable inductance systems form a second class of circuit choices. Their operating characteristics and degree of performance are highly dependent upon the way variability is built into the inductors. The number of tuned circuits is not always as important as the particular design used for the variable inductance and the manner in which they are used in the system. These are best treated on a case by case basis.

Single Tuned Receivers

These simple systems use one tuned circuit, as shown in Figure 3-3. Component values are for broadcast band applications, but these can be changed for other frequencies.

The circuit in Figure 3-3A tunes very broadly unless the antenna is deliberately shortened to just a few feet in length, and it seldom gives satisfactory results. One way of improving the poor selectivity is to use a large loop antenna. A six inch diameter or larger solenoid coil can also be used as a loop antenna if the receiver is close to a local station. Selectivity is good, but the received signal may be very faint. In any of the systems described here some improvement may be possible by using either a tapped coil or coil with a slider. Another alternative is to move the variable capacitor to the antenna circuit. This is often an improvement when a long antenna can be used.

Two coils are used in the Figures 3-3B-D and these circuits have a greater range of flexibility. By preadjusting the mutual inductance or by building variocoupling into the receiver coupling can be adjusted. Figure 3-3B works best with a long antenna. The set of figure 3-3C works with a short antenna, and also allows a longer antenna combined with variocoupling .

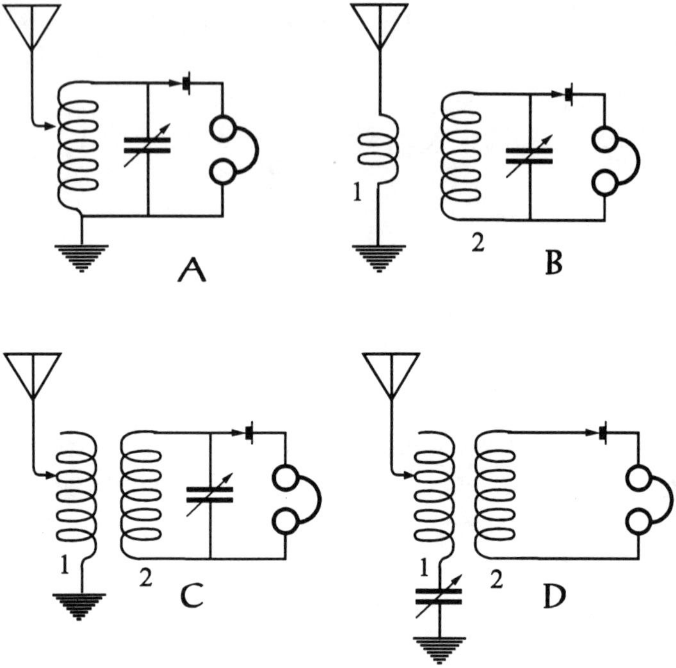

Figure 3-3: Receivers with one tuned circuit.

Coil and Variable Capacitor Notes:

Untuned antenna coil wound with few turns for low inductance.

Broadcast band coil. Typical inductance is 250µh.

Variable capacitor with nominal rating of 350 to 365pf.

One of the reasons single tuned circuits with a tuned detector show poor selectivity is due to the energy drawn directly from the resonant circuit to drive the earphones. This decreases the Q of the tuned LC circuit. The circuit in figure 3-3D can be an exception to this situation, if loose coupling is maintained between the detector coil and the tuned antenna.

Another example using an untuned detector is shown in Fig. 3-4C. Here a wave trap is coupled to the detector coil, which is loosely coupled to the antenna coil. The net action of the wave trap is to reject energy at the tuned frequency to the rest of the system, and this energy is then available at the detector, while signals at other frequencies are non-resonant in the system and have low amplitudes. Although the trap itself performs as a rejector, the system as a whole accepts only the desired signal. This approach has sometimes been called absorption tuning. Both mutual inductances can be established by variocoupling to make this receiver a versatile and remarkably efficient system. It can be expected to perform comparably with some receivers employing two tuned circuits which are described in the next section. This is also true of the system which results when the wave trap is inductively coupled to the antenna coil rather than the detector circuit. Experiments are recommended to see which is best for a given installation.

Figures 3-3B through 3-3D show one or more ungrounded (floating) circuits. Locations 1 and 2 in 3-3B, C, and D may to connected together and to ground on an experimental basis. Usually this improves sensitivity and decreases selectivity.

Capacitive coupling may be substituted for inductive coupling in any of these systems. However, there seems to be no particular advantage in using this type of coupling, and at least one early day reference indicated that broader tuning would result. Also, additional components are necessary. The same is true of more elaborate types of coupling circuits, which are discussed in electronic engineering texts. They require even more components than are needed for capacitive coupling. Nevertheless, experimentalists may find it interesting to investigate these possibilities.

The variable capacitor in circuits such as those of Fig. 3-3 can be replaced by a fixed unit, with tuning accomplished by a variometer type of tuning coil. This method was widely used in equipment built prior to the mid 1920s. Except for reproductions and replicas, there is no real need for this approach.

All of the systems with tuned antennas show the variable capacitor on the ground side of the antenna coil, which is good design practice. However, the broader tuning versions of the sets are so nonselective that hand capacity is not important. In such cases, the variable capacitor can be placed on the antenna side of the coil if desired. The change in resonant frequency caused by hand capacitance is not enough to produce a substantial change in signal level.

Most of the receivers in Fig. 3-3 use a minimum of parts, making them relatively simple to build. This can also lead to compact packaging as well as cost savings. Sometimes this becomes especially important when early day components are used. One variable capacitor is easier to find than a matched pair, for example. Tuning these sets is easy, and a beginner should try some of them just to get an idea of the limitations on crystal set performance.

Receivers with Two Tuned Circuits

Some receivers with two tuned circuits appear in Figure 3-4, and several are upgraded versions of the single tuned circuits discussed in the previous section. The set of Figure 3-4A usually shows a marked improvement in selectivity over those of Figures 3-3A and C, when variocoupling is also available. It is still a relatively simple set to build and operate, although it is intended only for a long antenna. Figure 3-4B is the very short antenna alternative. These sets can be effectively combined in one receiver using the switching circuit in Fig. 3-2. Absorption tuning is used in Figures 3-4C and D. Improvement may be relatively small, however, since power loss in the earphones lowers the Q of the detector circuit.

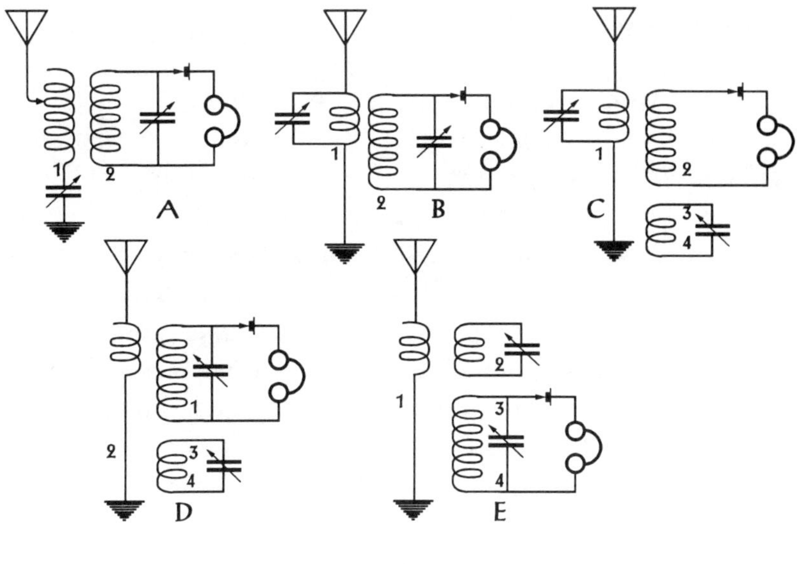

Figure 3-4: Receivers with two tuned circuits.

Fig. 3-4C works with a very short antenna when the antenna coil is tuned. Fig. 3-4D may be a better choice for an untuned antenna with intermediate length. The coils in Fig. 3-4E are arranged so that energy received from the antenna is transferred from the low inductance antenna coil, through the tuned intermediate circuit, to the tuned detector. There is no significant amount of coupling between the antenna and detector coils. However, the benefit of antenna tuning is not utilized here, so performance is limited and usage is suggested only where a tunable antenna cannot be employed.

The tapped coil and slider coil options, switching to tune either a long or short antenna, and the ground connection possibilities, all discussed for single tuned receivers, are also applicable here. Variocoupling is another desirable feature. Adding a tuned circuit results in the type of receiver covered in the next section.

Receivers with Three Tuned Circuits

Receivers with three tuned circuits were employed well before World War I, and were recommended as a desirable combination of three general methods of tuning. These systems show the high degree of sophistication achieved even before World War I. Most capacitors and inductances in these systems are variable, and this permits selecting optimum values of all components for best operation. There are disadvantages nonetheless. These are very complicated systems, so the operator must be highly skilled, familiar with the receiver, and acquainted with local receiving conditions. Also, the coils are not as efficient as might be supposed, because the unused open-ended portions lower the Q, due to the dead end effect. The latter is caused by self-induced voltages in these unused windings, with consequent losses.

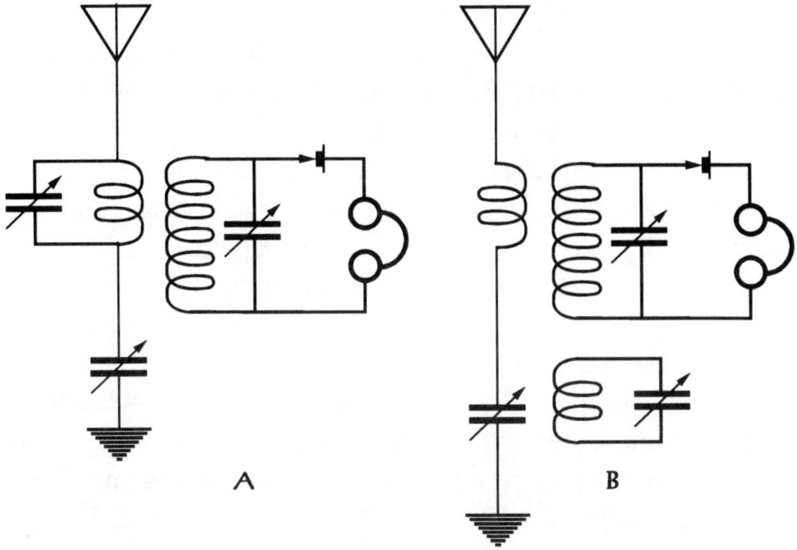

Figure 3-5: Receivers with three tuned circuits.

The arrangement in Fig. 3-5A is a much simplified version of the early day three tuned circuit receiver. It has two tuning capacitors in the antenna circuit, plus a tuned detector. The two antenna capacitors are strongly interactive. That is, changing the tuning of one makes it necessary to fully retune the other. The best combination of dial settings depends upon the length of the antenna and the signal frequency. This receiver is not easy to operate at its best, even though it is much less complicated than the early day version.

Fig. 3-5B is an improved version of the double tuned receiver in Fig. 3-4A, with absorption tuning added at the detector coil to increase selectivity. Sometimes a simple rearrangement of the coils will make a difference in performance.

The overall efficiency can be very high using three tuned circuits. Excellent selectivity is possible without the drastic sacrifice of sensitivity that is usually necessary with simpler systems. On the other hand, a relatively large number of components is required, as well as additional layout and construction work. Also, tuning three variable capacitors can become tedious, particularly with the degree of selectivity which can be reached with such receivers.

The antenna tuner circuit cannot be expected to track the other tuned circuits across the entire broadcast band, but with some sets the other two variable capacitors can be ganged together for simpler tuning. Alternatives for making common connections or connections to ground are reduced when a conventional two gang tuning capacitor is used, because of the metal rotor shaft common to both units.

Detector circuits generally track well with an absorption tuned circuit or energy transfer circuit, when a fixed tap is used on the detector coil. The two circuits can be adjusted for tracking by means of the trimmer capacitors usually provided with the two gang unit, or separate trimmers can be installed. Tracking may be poor when a multi-tapped high Q detector coil is used. This occurs because the wiring to the taps introduces additional stray capacitances around the detector coil, changing its tuning characteristics. The wiring needs to be adjusted only if a large change is made in the tuned frequency. A slider may be used instead of multiple taps with less effect upon tracking, and it is an attractive choice for this application.

High Q systems have another disadvantage that becomes very noticeable when large diameter air core coils are used. Stray capacitances and unwanted mutual inductances become so large that circuit characteristics are altered, usually in unpredictable and undesirable ways. These effects can become important when three inch or larger diameter coils are used, and they depend upon the size of other parts and the physical layout, as well as the particular circuit that is used. Ferrite core coils, with their combination of high Q and small size, can be used to advantage in these systems, because the unwanted side effects are greatly reduced. Unwanted mutual inductances also can be reduced by using closed field type binocular coils.

Variable Inductance Tuned Receivers

Many crystal sets, especially the early day models, tune by some method of changing inductance rather than by changing capacitance. Inductance is changed either by a series of taps from the coil to a switch, by a sliding conductive contact along the coil, or by a variometer. These possibilities were mentioned earlier. Combined variable inductance and variable capacitance tuning is also used in some cases. A great number of circuit combinations are possible, and in some cases performance may be more dependent upon the variable inductance design and circuitry than upon the number of tuned circuits. It is impractical to separately discuss all of these combinations, and they can be considered as variations of the receiver designs discussed earlier.

Wave Trap Applications

Circuits can be added to receivers in any of the categories covered earlier, either for overall improvement, or for a specific application. Acceptor wave traps were discussed earlier, because of their frequent use for absorption tuning in capacitance tuned circuits. However, rejector wave traps are used quite differently. One or more rejector wave traps can be used with any receiver, but this is considered a special application. A rejector trap affects receiver performance in the portion of the band near its tuned frequency rather than over the band as a whole. Nevertheless, the only practical use is to preset it and leave it. The receiver tunes over most of its frequency range as if the trap were not present.

As stated in Chapter two, the most commonly chosen location for a rejector wave trap is in the antenna circuit. Rejection is at least as efficient in the antenna circuit as at any other location, and there is less disruption of normal tuning at nearby frequencies. The opinion of some crystal set enthusiasts is that it is better to improve overall receiver selectivity by other methods than to use a wave trap to minimize an interference problem. Nevertheless, a rejector trap may be worthy of consideration for an unusual special case, such as one very powerful local station which is a major cause of interference. Wave traps of either type can be added to the circuits described.

BIBLIOGRAPHY

Anderson, John E., Arthur C.C. Mills, and Elmer H. Lewis, Henley's 222 Radio Circuit Designs. Korman W. Henley Publishing Co., New York, 1923, p.67. Reprinted by Lindsay Publications, Inc., Bradley IL, 1989.

Anderson, Phil, The Miller '595' Tuner Revisited. The Xtal Set Society Newsletter. Vol. 2, No. 5, May. 1993, p. 1.

Bucher, Elmer E., Practical Wireless Telegraphy. Wireless Press, Inc., New York, 1917, p. 144.

Dynner, Eugene, An Exceptional Radio Receiver. Radio Amateur News, Vol.1, No.2, Aug. 1910, p. 63.

Gardner, Leonard U., W2QBC, 458 Two Mile Greek Rd., Tonawanda, NY 14150, personal communications, Jan. 31, 1991 and Jan. 31, 1993.

Harrison, Arthur, Constructing Wireless-Era Receivers from Authentic Components. Antique Radio Classified, Vol. 9, No. 11, Nov. 1992, p.18.

Hayward, James, At Last, a Portable Crystal Set. The Xtal Set Society Newsletter, Vol. 4, No. 1, Jan. 1, 1924, p.2.

Horvath, Joe, Uncle Joe's No. 4 Super Selective Single Dial Control Crystal Set. California Antique Radio Gazette, Vol. 11, No. 2, May 1986, p.14.

Horvath, Joe, Coil Data for Uncle Joe's Super Selective Tuner or Crystal Set. Op. Cit., Vol. 12, No. 1, Feb. 1987, p. 23.

Kinzie, P.A., Some Experiments with Crystal Sets Using High Q Coils. Arizona Antique Radio Club News, Vol. 7, No. 1, Spring 1990, p. 12.

Pierce, George W., Principles of Wireless Telegraphy. McGraw-Hill Book Co., Inc., New York, 1910, p. 286.

Osterhoudt, Elmer G., MRL 13 Crystal Set Circuits. Handbook HB-25. Modern Radio Laboratories, 1947, Minneapolis, MN.

Osterhoudt, Elmer G., MRL Detail Print No. DP22.

Osterhoudt, Elmer G., MRL Detail Print No. DP23.

Osterhoudt, Elmer G., MRL Detail Print No. DP32.

Strozier, Chuck, A Crystal Set. California Antique Radio Gazette, Vol 15, No. 4, Nov. 1990, p. 18.

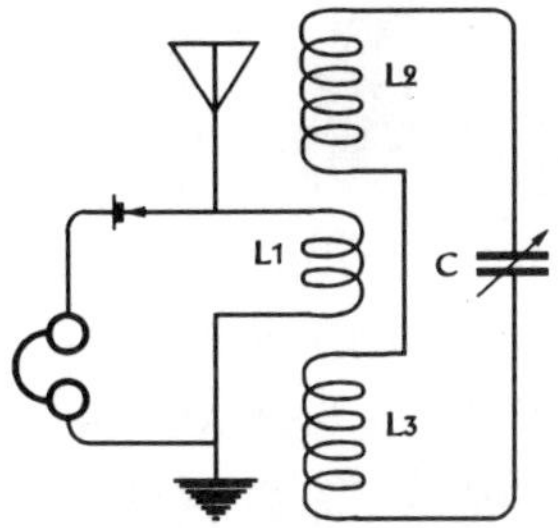

CHAPTER 4

CRYSTALS FOR DETECTORS

Materials for Detector Applications

Most active substances which have been used in crystal sets are crystalline materials which originated either in ore deposits or as synthetic manufactured products. As usual, there are exceptions to the rule, such as the fused silicon product of the electric furnace, which can be non-crystalline. The intention here is to discuss the practical application of these materials rather than to precisely categorize. Therefore, if it looks like a detector material, acts like a detector material, and is used like a detector, it is included.

It is surprising how few unusual cases are included, using this selection rule, in view of the lack of restrictions in the statement. Two of these are electrolytic detectors, using lead peroxide or zinc peroxide. They were used in the form of compressed powder pellets. A few other materials depend at least in part on the principle of imperfect contact rectification. The carbon-steel detector qualifies here, the extreme example being the "fox hole" radio of World War II. Its detector was a piece of pencil lead resting across two parallel razor blade edges.

There are some less clear-cut examples of active substances showing imperfect contact behavior. Some samples of one or two crystalline compounds have been reported to improve in sensitivity as the surface becomes tarnished, when a catwhisker contact is used. One possible explanation involves imperfect contact supplementing the usual solid state rectification property. It should be noted, however, that this behavior is very rare. Performance usually deteriorates as the surface of a crystal oxidizes.

There are no firm rules regarding the composition and appearance of a substance which might establish it as a good solid-state rectifier. There are only trends that suggest the possibility. Early day investigators were hampered and sometimes misled by this state of confusion; a satisfactory theory did not exist to explain rectification. Early work was entirely empirical, including the first demonstration of solid-state rectification by Ferdinand Braun. Braun noted some of the trends that led to good examples, being first to point out that metal sulfides are noteworthy in this respect.

Later investigators noticed a certain degree of correlation between rectification and the thermoelectric properties for many intermetallic compounds, such as tellurium, which is electrically conductive to some extent. The correlation was so noticeable that for a time it was widely believed that thermoelectric effects caused rectification due to the heat presumably generated at the point contact. It required some very difficult laboratory experimentation to disprove this theory.

The appearance of a material sometimes gives a clue about its performance, but this test is inconclusive. Most good detectors have a shiny metallic appearance ranging from silvery white through steel gray. Golden and coppery crystals also occur. Nevertheless, a few detectors have a dull nonmetallic appearance.

There are other material properties that have important effects on the preparation and use of an active substance as a crystal detector. Heat sensitivity is a problem with some materials, including galena, one of the most used crystals. Heat may either accelerate surface oxidation or alter the bulk material, and this degrades sensitivity. The sensitivity of galena and certain other crystals may be destroyed locally at the point contact by static discharges or even by a very strong signal, all because of the sensitivity to heat that is generated at the contact. Mounting such crystals in a metal base requires special care, and this will be discussed in the next section.

Most crystal surfaces are all too easily damaged by careless handling. Your fingers should not touch an active surface. It may be possible to clean a freshly contaminated crystal with methyl ethyl ketone (PEK) or acetone, but no great degree of success has been reported in this respect. One exception, zincite has been reported to be cleanable with carbon disulfide after becoming contaminated. On a longer term basis, most materials exhibit a progressive deterioration simply through exposure to air. Tarnish caused by surface oxidation is the causative factor. The rate of oxidation varies with composition. Some types of crystals need to be replaced after a few months. With others the useful lifetime may be measured in years. In any case, heat speeds up the deterioration process.

Typical detector materials are relatively hard and brittle. Even so, just the repeated adjustment of a catwhisker may eventually mar the surface, resulting in decreased performance. The use of sharp pointed contacts or the broad end contact of a second crystal can easily destroy the surface as far as sensitivity is concerned. Extreme care must be taken while making adjustments to these types of connections.

There are almost no exceptions to the rule that attempts to re-surface a crystal, by polishing, sanding, or grinding, will destroy it. The noteworthy exception is molybdenite. This compound can actually be restored to high sensitivity by sanding the exposed surface. In practically all other cases, the only alternative to replacement is to split the crystal and remount it with a fresh surface exposed. Even this procedure does not work if the interior has been spoiled by overheating.

Up to this point, materials properties have been discussed only in a generalized way. Tens of thousands of different compositions and combinations of materials have individual characteristics that have been tested for rectification over the years, and most results were never made public. It can be assumed that the reason for such lack of information is simply that performance was poor, so there was no incentive to report the results.

Contacting Methods

Point Contact Junctions

The first successful point contact method for practical crystal detectors depended upon a stiff, pointed wire, or even a needle or pointed rod. Contact pressure varied, depending upon the crystal. In some cases this provided a fairly stable detector for rugged applications onboard ships and in army field equipment. The most sensitive and yet practical materials with this type of contact were carborundum (silicon carbide), cerussite (lead carbonate), and molybdenite (molybdenum disulfide). Silicon was also used to a degree. This was the fused silicon electric furnace product, not today's highly purified, doped silicon crystals.

The light wire contact method, which includes the very delicate catwhisker just touching the active substance surface, evolved later. The catwhisker contact was patented by Greenleaf Whittier Pickard, one of the inventors of the crystal detector itself, but only after several years had passed following his first patent. Pickard later remarked that the catwhisker must have been independently invented and used by others before he filed for his patent.

Initially, there was a lack of interest in the catwhisker, because it is so susceptible to vibration and shock. However, its performance in a benign environment is unequaled by other methods, when a crystal of the proper composition is used. It became the most used contacting method when the advent of broadcast radio provided a demand for high sensitivity without the pressing need for reception in rough environments. This brought about a somewhat different choice of crystals, with compositions for use in the home. The materials with the highest sensitivity were steel galena, galena, iron pyrites, and fused silicon. At present, the germanium diode is the fixed crystal substitute for the galena materials. It has equal or nearly equal sensitivity, and there is no requirement for adjustment.

A comparison of the sensitivity and performance of pointed wire contact versus catwhisker contact is difficult to evaluate when there is little knowledge of which method was used with crystals. For example, textbooks published before about 1912 usually discuss sensitivity in terms of pointed wire or rod contact, and list the most sensitive crystal as cerussite or molybdenite, with carborundum also included. In contrast, once the catwhisker came into use, textbooks listed galena and iron pyrites as sensitive. The composition of the catwhisker wire itself also appears to make some difference in performance. However, there is a lack of agreement concerning which metal is best, because in some cases the choice of crystal affects the choice of catwhisker metal, which has complicated the acquisition of data.

There is another contact method, not previously mentioned, and which also preceded the catwhisker. This was the use of a second active substance, usually in the form of a pointed crystal, held in contact with the surface of the first under moderate pressure. The use of two crystals for detection was another Pickard invention, and it is commonly referred to as a Perikon. This is a generalization of Pickard's original name for a specific pair of materials. Like a pointed rod under pressure, a Perikon detector was suitable for rugged environments and was used to some degree for shipboard installations. However its sensitivity did not equal that of a catwhisker.

Other Contacting Methods

In modern fixed diodes no fully satisfactory substitute exists for a point contact junction, established either by mechanical pressure, or by a delicate spot weld. Results with other methods have been partially successful at best. The earliest carborundum detector is the classic example. It consisted of a relatively long carborundum crystal wrapped with copper wire at each end. The rejection rate during the first production run was on the order of 90 percent. It was later concluded that rectification in the better detectors was due to a pronounced bulk properties change along the length of the crystal. A relatively uniform crystal detected poorly.

A few other active substances have been used by mounting them between two plates under pressure. Different plate metals were sometimes used to improve detection. The shape of the contacts was a factor in some cases; at least two patents were issued in the 1900-1910 period which described a cylinder of metal contacting the active surface of the material. None of these methods equaled the sensitivity of a fine wire point contact junction.

A closer approach to satisfactory operation was obtained through the use of either a powder or a brush of many fine wires as a substitute for a catwhisker. Enclosed, even scaled cartridges, were sold during the first years of the broadcast era, beginning in the early 1920's. Usually the sales pitch emphasized that many separate contacts eliminated the problem of a sudden loss of contact — which was characteristic of a disturbed catwhisker — as well as the frustrating search for a sensitive spot. Some of these products provided for user adjustment "just in case."

The Base Connection

The best form of base connection for point contact detectors consists of a good wide area electrical contact. This can be produced by casting the active substance, partially immersed, in a short metal cup with some low melting point metal. Acceptable performance is usually possible if the specimen is placed in a hollow holder and a soft metal wool or foil is tightly packed around it. Fine steel wool has enough of a packing characteristic to be used in the way.

Application Technique

Choosing a Crystal

At present the selection of a crystal for new construction projects is relatively simple, as far as choosing from existing products is concerned. These include solid state diodes as well a premounted crystals. The 1N34 germanium diode gives good performance along with the advantages of constant sensitivity and no requirement for adjustments. Other diodes that have been used and recommended include the 1N21, 1N22, 1N60, 1N66, 1N69, 1N128, 1N294, and 1N295. Silicon diodes work well but have no particular advantages, and they require a small amount of bias voltage for best results.

The property of constant sensitivity is highly desirable when the object is to evaluate other components or circuit changes. There is no added variable to confuse the results. In other words, if circuit B works better that A, there is no likelihood that change was caused by the detector instead of the circuit. This may occur when using a catwhisker crystal. However, when sensitivity is all-important, a galena crystal works least as well as a 1N34, and occasionally even better. Some steel galena crystals have the same level of sensitivity. Iron pyrites are not quite as sensitive as the galena types for the broadcast band, but they are the preferred choice for short wave receivers. Silicon is satisfactory for the broadcast band, but carborundum is definitely less sensitive, even with the proper bias voltage applied.

The picture is quite different for the case of investigating crystals themselves. A study of modern synthetics, which were unavailable to early day experimenters, is an interesting field for those who like to be among the first to try something new. There is also the possibility of finding good catwhisker crystals among the materials which were originally used with heavy contact methods and were discarded. Also, new sources of ore have been found since the early wireless period, and they may have better properties than those previously used. In short, there are a large number of possible sources to choose from, but locating the specimens can be a limitation.

Different limitations are associated with restoration projects or with construction of a reproduction. This is because most equipment manufactured before the end of World War I did not use a catwhisker crystal for detection. Therefore, it would be inappropriate to choose a galena crystal and catwhisker for such a project without supporting documentation, even though the method was well known by that time. The only certain way to make the right choice is to determine what was used in the specific receiver.

Some general guidelines are available when specific information cannot be found and typical usage is an acceptable criterion. Pre-World War I American Marconi receivers used carborundum crystals for several years, along with heavy pointed contacts under considerable pressure. The parent Marconi company in England adapted the same type of detector after retiring earlier detection methods. Somewhat later, both carborundum and cerussite crystals were used in dual holders that allowed an operator to switch from one to the other. Silicon was used on detector stands sold by the Wireless Specialties Products Company. These were installed in a number of different receivers. Molybdenite was also available, and there were Perikon type combinations with two different crystals also in use. All of these were of the heavy contact junction type.

Wireless operators on ships sometimes carried their own favorite crystals with them, regardless of company policy discouraging that practice. Amateurs were noted for skill and interest in experimenting. Therefore, it is likely that a wide variety of materials actually received at least some usage. As stated earlier, the catwhisker technique was also known, although the most likely users were amateurs and shore station operators in a relatively quiet environment. When evidence of such usage on a specific set exists, it is reasonable to restore it for the same less-than-typical choice of detector.

There were factors which affected the choice of a crystal during the early wireless era which are not obvious under more modern conditions. Some understanding of these is helpful when working with early day designs. First, carborundum was often the crystal of choice because of its ruggednesss and stability, even though it was less sensitive that the other commonly used materials. Its tolerance of heat and high currents allowed it to be used where current surges that damaged materials more susceptible to heating were apt to occur. Static discharges from the long, high, shipboard antennas then in use were one source of surges. Further, by necessity, shipboard transmitters and receivers either used the same antenna or closely adjacent lengths of wire, and were located close together. In spite of precautions to avoid high currents through the detector, there was a problem with the less heat resistant materials. In fact, a strong local signal from another ship or nearby shore station was sufficient to cause damage in some cases. Carborundum could survive under such conditions.

The hardness associated with carborundum was another advantage. A stiff wire, needle, or pointed rod could be held against the surface of the crystal under considerable pressure without destroying a sensitive spot. High pressure greatly decreased the chances of vibration or shock causing contact point movement and loss of sensitivity. Other materials were not as tolerant of contact point pressure, and thus they were more prone to being knocked out of adjustment.

The fused silicon electric furnace product then available was considered to be generally satisfactory, although somewhat less rugged than carborundum. Higher sensitivity was a compensating factor. Cerussite and molybdenite are softer than either carborundum or silicon, so lighter contact pressure was necessary, leading to more problems with vibration and shock. They are also more sensitive to heating, but when these properties could be accepted, the resulting sensitivity was higher than either carborundum or silicon. Galena was well down on the list of crystals because of its poor heat sensitivity and rapid surface oxidation characteristics. When used with a heavy contact it was not as sensitive as cerussite or molybdenite. Other minerals which were tried had similar disadvantages and were little used.

BIBLIOGRAPHY

Constable, Anthony, Early Wireless. Midas Books, 12 Dene Way, Speldhurst, Tunbridge Wells, Kent, England, 1980.

Gibbons, Robert C., (Editor), Woldman's Engineering Alloys, Sixth Edition. American Society for Metal, Metals Park, Ohio, 1979.

Knoll, Max, Materials and Processes of Electron Devices. Springer-Verlag, Berlin, 1959.

Kohl, Walter H., Handbook of Materials and Techniques for Vacuum Devices. Reinhold Publishing Corp., New York, 1957, p. 358.

Osterhoudt, Elmer G., Crystal Detectors. MRL Handbook HB-3, Modern Radio Laboratories, Minneapolis, MN 1954.

Osterhoudt, Elmer G., Facts for Crystal Experimenters. MRL Handbook H9-10, Modern Radio Laboratories, Minneapolis, MN, 1954.

Sievers, Maurice L., Crystal Clear Volume I. The Vestal Press, Ltd., Vestal, New York, 1991.

ABOUT THE AUTHOR

P.A. (Phil) Kinzie earned a Master's Degree in Engineering at UCLA, and worked with aircraft ground and flight test instrumentation for many years, with later similar employment in the space program. More recently he was registered as a Control Systems Engineer in California, with experience in oil refinery and similar industrial projects.

Although not currently in engineering, he has continued a long-time interest in crystal sets that began at the age of nine or ten. At that time he lost his enthusiasm for dismantling broken clocks and watches, and turned his attention to old radios. He dissected these to extremes that included unwinding the paper-foil capacitors to see what was inside.

Destruction is not usually the recommended way of entering a new field, but in his case he eventually graduated to a better pastime of building crystal sets and one tubers, a hobby that he still continues when time permits. This book reflects his effort to analyze and improve upon the special technology needed for design and construction of crystal sets.

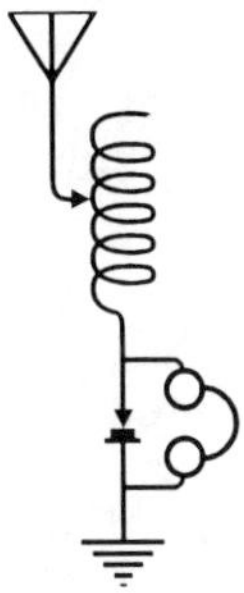

THE XTAL SET SOCIETY

The Xtal Set Society ***Newsletter***, bi-monthly, one year subscription. (postage is included, no additional shipping charge) $12.95
International subscriptions please remit US$19.00
Canadians please remit US US$14.00

Volume I & II of the Society Newsletter, twelve issues, May, 1991 through May 1993. $ 19.95

The Crystal Set Handbook and Volume III of the Society Newsletter, three issues, ending Nov. 1993. $ 11.95

Volume IV of the Society Newsletter, six issues, ending November 1994. $ 10.95

Crystal Sets: Volume V of the Society Newsletter, six issues, ending November 1995. $ 10.95

Crystal Set Projects: 15 Radio Projects You Can Build
New book! Written by members of the Society, 1997 $ 14.95

Crystal Set Building and More: Volume 6 & 7 of the Society Newsletter, 12 issues ending November 1997 $ 15.95

Crystal Set Loopers, a Three Tuber & More: Volume 8 of the Society newsletter, 6 issues ending November 1998 $ 14.95

Shipping and handling on book orders $ 3.50

The Xtal Set Society
www.midnightscience.com
e-mail: xtalset@midnightscience.com
1-800-927-1771 VISA/MC accepted

Crystal Set Projects: 15 Radio Projects You Can Build

Crystal Set Projects is a collection of radio projects that won a recent building contest run by the Society. Thanks to everyone who participated. If you missed out, don't worry. We're doing another! Included are step-by-step instructions so you can build and design your own Xtal sets no matter what your background in electronics. The projects are of various designs and difficulty so that everyone can learn something. "Crystal set circuits have been regularly published in the hobby press since the dawn of radio. But the Xtal Set Society's new project book covers more electronic and mechanical variations of this ever-popular receiver than I have ever seen in one place. You'll find hi tech and low tech versions, vintage and modern versions, sets with loop antennas and short-wave bands. The book is as much fun to read as to build from, but I defy anyone to browse this infectious publication for long without experiencing an uncontrollable urge to break out the soldering iron and double-cotton-covered wire!" —Marc Ellis, Editor, The Old-Timer's Bulletin of the AWA; Contributing Editor, Antique Radio, Popular Electronics magazine. Get yourself a copy! 6 x 9 paperback, 160 pages, $14.95.

The Xtal Set Society is dedicated to once again building and experimenting with radio electronics, often—but not always—through the use of the crystal set, the basis for most modern day radio apparatus. The Society newsletter helps support our goal of producing excellent quality technical books that encourage learning and building. To join the society and receive one year (6 issues) of the bi-monthly newsletter, remit $12.95 to The Xtal Set Society. Canadians, please remit US $14.00. Outside the US and Canada please remit US $19.00.

The Xtal Set Society

1-800-927-1771 VISA/MC

e-mail: xtalset@midnightscience.com

www.midnightscience.com

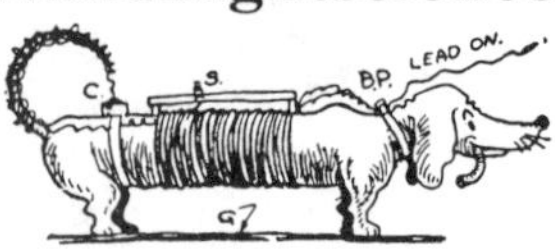